STUDENT WORKBOOK

PHYSICS

SCIENTISTS AND ENGINEERS **A STRATEGIC APPROACH** 4/E

WITH MODERN PHYSICS

NDALL D. KNIGHT

PEARSON

Editor-in-Chief: Jeanne Zalesky

Acquisitions Editor: Darien Estes

Project Manager: Martha Steele

Program Manager: Katie Conley

Senior Development Editor: Alice Houston

Development Manager: Cathy Murphy

Program/Project Management Team Lead: Kristen Flathman

Production Management: Anju Joshi

Compositor: Lumina Datamatics®

Main Text Cover Designer: John Walker

Supplement Cover Designer: Seventeenth Street Studios

Rights & Permissions Project Manager: Maya Gomez

Rights & Permissions Management: Rachel Youdelman

Manufacturing Buyer: Maura Zaldivar-Garcia

Executive Marketing Manager: Christy Lesko

Marketing Manager: Elizabeth Ellsworth

Cover Photo Credit: Thomas Vogel/Getty Images

ISBN 10: 0-134-08316-4; ISBN 13: 978-0-134-08316-2

Table of Contents

Part VI — Electricity and Magnetism

Part VII — Optics

Part VIII — Relativity and Quantum Physics

Dynamics Worksheets
Energy Worksheets
Momentum Worksheets

Preface

Learning physics, just as learning any skill, requires regular practice of the basic techniques. That is what this *Student Workbook* is all about. The workbook consists of exercises that give you an opportunity to practice the ideas and techniques presented in the textbook and in class. These exercises are intended to be done on a daily basis, right after the topics have been discussed in class and are still fresh in your mind.

You will find that the exercises are nearly all *qualitative* rather than *quantitative*. They ask you to draw pictures, interpret graphs, use ratios, write short explanations, or provide other answers that do not involve significant calculations. The purpose of these exercises is to help you develop the basic thinking tools you'll later need for quantitative problem solving. Successful completion of the workbook exercises will prepare you to tackle the more quantitative end-of-chapter homework problems in the textbook. It is highly recommended that you do the workbook exercises *before* starting the end-of-chapter problems.

You will find that the exercises in this workbook are keyed to specific sections of the textbook in order to let you practice the new ideas introduced in that section. You should keep the text beside you as you work and refer to it often. You will usually find Tactics Boxes, figures, or examples in the textbook that are directly relevant to the exercises. When asked to draw figures or diagrams, you should attempt to draw them so that they look much like the figures and diagrams in the textbook.

Because the exercises go with specific sections of the text, you should answer them on the basis of information presented in *just* that section (and prior sections). You may have learned new ideas in Section 7 of a chapter, but you should not use those ideas when answering questions from Section 4. There will be ample opportunity in the Section 7 exercises to use that information there.

You will need a few "tools" to complete the exercises. Many of the exercises will ask you to *color code* your answers by drawing some items in black, others in red, and yet others in blue. You need to purchase a few colored pencils to do this. The author highly recommends that you work in pencil, rather than ink, so that you can easily erase. Few people produce work so free from errors that they can work in ink! In addition, you'll find that a small, easily carried six-inch ruler will come in handy for drawings and graphs.

As you work your way through the textbook and this workbook, you will find that physics is a way of *thinking* about how the world works and why things happen as they do. We will be interested primarily in finding relationships and seeking explanations, only secondarily in computing numerical answers. In many ways, the thinking tools developed in this workbook are what the course is all about. If you take the time to do these exercises regularly and to review the answers, in whatever form your instructor provides them, you will be well on your way to success in physics.

To the instructor: The exercises in this workbook can be used in many ways. You can have students work on some exercises in class as part of an active-learning strategy. Or you can do the same in recitation sections or laboratories. This approach allows you to discuss the answers immediately, to answer student questions, and to improvise follow-up exercises when needed. Having the students work in small groups (2 to 4 students) is highly recommended.

Alternatively, the exercises can be assigned as homework. The pages are perforated for easy tearout, and the page breaks are in logical places so that you can assign the sections of a chapter that you would likely cover in one day of class. Exercises should be assigned immediately after presenting the relevant information in class and should be due at the beginning of the next class. Collecting them at the beginning of class, then going over two or three that are likely to cause difficulty, is an effective means of quickly reviewing major concepts from the previous class and launching a new discussion.

If the exercises are used as homework, it is *essential* for students to receive *prompt* feedback. Ideally this would occur by having the exercises graded, with written comments, and returned at the next class meeting. Posting the answers on a course website also works. Lack of prompt feedback can negate much of the value of these exercises. Placing similar qualitative/graphical questions on quizzes and exams, and telling students at the beginning of the term that you will do so, encourages students to take the exercises seriously and to check the answers.

The author has been successful with assigning *all* exercises in the workbook as homework, collecting and grading them every day through Chapter 4, then collecting and grading them on about one-third of subsequent days on a random basis. Student feedback from end-of-term questionnaires reveals three prevalent attitudes toward the workbook exercises:

i. They think it is an unreasonable amount of work.
ii. They agree that the assignments force them to keep up and not get behind.
iii. They recognize, by the end of the term, that the workbook is a valuable learning tool.

However you choose to use these exercises, they will significantly strengthen your students' conceptual understanding of physics.

Following the workbook exercises are optional Dynamics Worksheets, Energy Worksheets, and Momentum Worksheets for use with end-of-chapter problems in Parts I and II of the textbook. Their use is recommended to help students acquire good problem-solving habits early in the course. A few blank worksheets, which you can have students photocopy, are at the back of the *Student Workbook* for the Extended and Standard editions and for Volume 1. You can also download PDFs of the worksheets via the "Resources" tab in the textbook's Instructor Resource Center (www.pearsonhighered.com/educator/catalog/index.page) or from the textbook's Instructor Resource Area in MasteringPhysics® (www.masteringphysics.com).

Answers to all workbook exercises are provided as pdf files and can be downloaded from the Instructor Resource Center or from the Instructor Resource Area in MasteringPhysics. The author gratefully acknowledges the careful work of answer writer John Filaseta of Northern Kentucky University.

Acknowledgments: Many thanks to Martha Steele at Pearson Education and to Anju Joshi at Lumina Datamatics® for handling the logistics and production of the *Student Workbook*.

1 Concepts of Motion

1.1 Motion Diagrams

1.2 Models and Modeling

Exercises 1–5: Draw a motion diagram for each motion described below.
- Use the particle model to represent the object as a particle.
- Six to eight dots are appropriate for most motion diagrams.
- Number the positions in order, as shown in Figure 1.4 in the text.
- Be neat and accurate!

1. A car accelerates forward from a stop sign. It eventually reaches a steady speed of 45 mph.

2. An elevator starts from rest at the 100th floor of the Empire State Building and descends, with no stops, until coming to rest on the ground floor. (Draw this one *vertically* since the motion is vertical.)

3. A skier starts *from rest* at the top of a 30° snow-covered slope and steadily speeds up as she skies to the bottom. (Orient your diagram as seen from the *side*. Label the 30° angle.)

4. The space shuttle orbits the earth in a circular orbit, completing one revolution in 90 minutes.

5. Bob throws a ball at an upward 45° angle from a third-story balcony. The ball lands on the ground below.

Exercises 6–9: For each motion diagram, write a short description of the motion of an object that will match the diagram. Your descriptions should name *specific* objects and be phrased similarly to the descriptions of Exercises 1 to 5. Note the axis labels on Exercises 8 and 9.

6.

8.

7.

9.

1.3 Position, Time, and Displacement

10. The figure below shows the location of an object at three successive instants of time.

    a. Use a **red** pencil to draw and label on the figure the three position vectors $\vec{r}_0$, $\vec{r}_1$, and $\vec{r}_2$ at times 0, 1, and 2.
    b. Use a **blue** or **green** pencil to draw a possible trajectory from 0 to 1 to 2.
    c. Use a **black** pencil to draw the displacement vector $\Delta\vec{r}$ from the initial to the final position.

11. In Exercise 10, is the object's displacement equal to the distance the object travels? Explain.

12. Redraw your motion diagrams from Exercises 1 to 4 in the space below. Then add and label the displacement vectors $\Delta\vec{r}$ on each diagram.

1.4 Velocity

13. The figure below shows the positions of a moving object in three successive frames of a video. Draw and label the velocity vector $\vec{v}_0$ for the motion from 0 to 1 and the vector $\vec{v}_1$ for the motion from 1 to 2.

Exercises 14–20: Draw a motion diagram for each motion described below.
- Use the particle model.
- Show and label the *velocity* vectors.

14. A rocket-powered car on a test track accelerates from rest to a high speed, then coasts at constant speed after running out of fuel. Draw a dashed line across your diagram to indicate the point at which the car runs out of fuel.

15. Galileo drops a ball from the Leaning Tower of Pisa. Consider the ball's motion from the moment it leaves his hand until a microsecond before it hits the ground. Your diagram should be vertical.

16. An elevator starts from rest at the ground floor. It accelerates upward for a short time, then moves with constant speed, and finally brakes to a halt at the tenth floor. Draw dashed lines across your diagram to indicate where the acceleration stops and where the braking begins. You'll need 10 or 12 points to indicate the motion clearly.

17. A bowling ball being returned from the pin area to the bowler starts out rolling at a constant speed. It then goes up a ramp and exits onto a level section at very low speed. You'll need 10 or 12 points to indicate the motion clearly.

18. A track star runs once around a running track at constant speed. The track has straight sides and semicircular ends. Use a bird's-eye view looking down on the track. Use about 20 points for your motion diagram.

19. A car is parked on a hill. The brakes fail, and the car rolls down the hill with an ever-increasing speed. At the bottom of the hill it runs into a thick hedge and gently comes to a halt.

20. Andy is standing on the street. Bob is standing on the second-floor balcony of their apartment, about 30 feet back from the street. Andy throws a baseball to Bob. Consider the ball's motion from the moment it leaves Andy's hand until a microsecond before Bob catches it.

1.5 Linear Acceleration

Note: Beginning with this section, and for future motion diagrams, you will "color code" the vectors. Draw velocity vectors **black** and acceleration vectors **red**.

Exercises 21–24: The figures below show an object's position in three successive frames of a video. The object is moving in the direction $0 \rightarrow 1 \rightarrow 2$. For each diagram:
- Draw and label the initial and final velocity vectors $\vec{v}_0$ and $\vec{v}_1$. Use **black**.
- Use the steps of Tactics Box 1.3 to find the change in velocity $\Delta \vec{v}$.
- Draw and label $\vec{a}$ next to dot 1 on the motion diagram. Use **red**.
- Determine whether the object is speeding up, slowing down, or moving at a constant speed. Write your answer beside the diagram.

21.
    2   1   0

22.
    2   1   0

23.
          2

        1

  0

24.
0

1

2

Exercises 25–29: Draw a complete motion diagram for each of the following.
- Draw and label the velocity vectors $\vec{v}$. Use **black**.
- Draw and label the acceleration vectors $\vec{a}$. Use **red**.

25. Galileo drops a ball from the Leaning Tower of Pisa. Consider its motion from the moment it leaves his hand until a microsecond before it hits the ground.

26. Trish is driving her car at a steady 30 mph when a small furry creature runs into the road in front of her. She hits the brakes and skids to a stop. Show her motion from 2 seconds before she starts braking until she comes to a complete stop.

27. A ball rolls up a smooth board tilted at a 30° angle. Then it rolls back to its starting position.

28. A bowling ball being returned from the pin area to the bowler rolls at a constant speed, then up a ramp, and finally exits onto a level section at very low speed.

29. Two sprinters, Cynthia and Diane, start side by side. Diane has run only 80 m when Cynthia crosses the finish line of the 100 m dash.

1.6 Motion in One Dimension

1.7 Solving Problems in Physics

30. The four motion diagrams below show an initial point 0 and a final point 1. A pictorial representation would define the five symbols: x_0, x_1, v_{0x}, v_{1x}, and a_x for horizontal motion and equivalent symbols with y for vertical motion. Determine whether each of these quantities is positive, negative, or zero. Give your answer by writing $+$, $-$, or 0 in the table below.

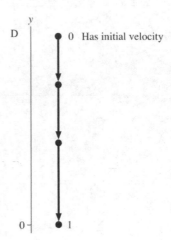

	A	B	C	D
x_0 or y_0				
x_1 or y_1				
v_{0x} or v_{0y}				
v_{1x} or v_{1y}				
a_x or a_y				

31. The three symbols x, v_x, and a_x have eight possible combinations of *signs*. For example, one combination is $(x, v_x, a_x) = (+, -, +)$.

    a. List all eight combinations of signs for x, v_x, a_x.

    1. _____   5. _____

    2. _____   6. _____

    3. _____   7. _____

    4. _____   8. _____

b. For each of the eight combinations of signs you identified in part a:
- Draw a four-dot motion diagram of an object that has these signs for x, v_x, and a_x.
- Draw the diagram *above* the axis whose number corresponds to part a.
- Use **black** and **red** for your $\vec{v}$ and $\vec{a}$. vectors. Be sure to label the vectors.

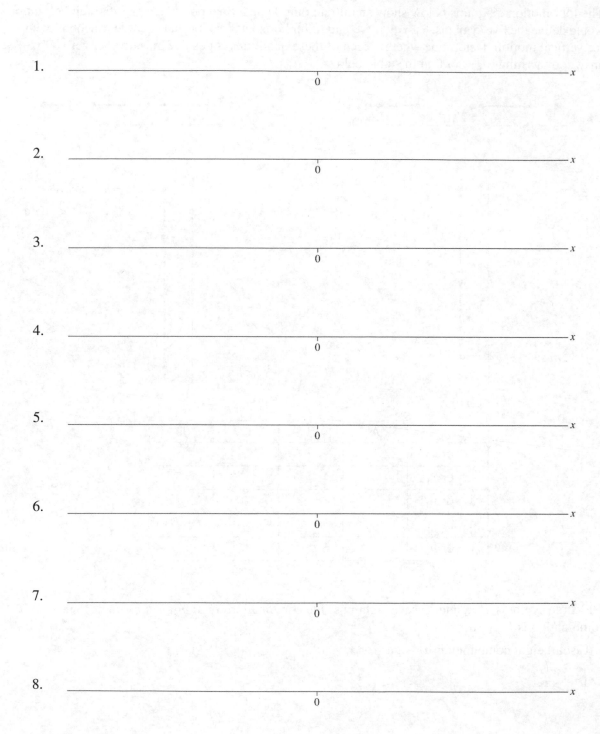

1. _____ x
   0

2. _____ x
   0

3. _____ x
   0

4. _____ x
   0

5. _____ x
   0

6. _____ x
   0

7. _____ x
   0

8. _____ x
   0

32. Sketch position-versus-time graphs for the following motions. Include a numerical scale on both axes with units that are *reasonable* for this motion. Some numerical information is given in the problem, but for other quantities make reasonable estimates.

    Note: A *sketched* graph simply means hand-drawn, rather than carefully measured and laid out with a ruler. But a sketch should still be neat and as accurate as is feasible by hand. It also should include labeled axes and, if appropriate, tick-marks and numerical scales along the axes.

    a. A student walks to the bus stop, waits for the bus, then rides to campus. Assume that all the motion is along a straight street and that the bus goes in the direction the student had been walking.

    b. A student walks slowly to the bus stop, realizes he forgot his paper that is due, and *quickly* walks home to get it.

    c. The quarterback drops back 10 yards from the line of scrimmage, then throws a pass 20 yards to the receiver, who catches it and sprints 20 yards to the goal. Draw your graph for the *football*. Think carefully about what the slopes of the lines should be.

33. Interpret the following position-versus-time graphs by writing a very short "story" of what is happening. Be creative! Have characters and situations! Simply saying that "a car moves 100 meters to the right" doesn't qualify as a story. Your stories should make *specific reference* to information you obtain from the graphs, such as distances moved or time elapsed.

   a. Moving car

   b. Submarine

   c. Two football players

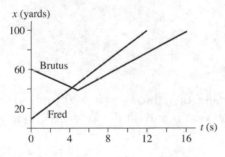

34. Can you give an interpretation to this position-versus-time graph? If so, then do so. If not, why not?

1.8 Units and Significant Figures

35. Convert the following to SI units. Work across the line and show all steps in the conversion.

    a. 9.12 μs ×

    b. 3.42 km ×

    c. 44 cm/ms ×

    d. 80 km/h ×

    e. 60 mph ×

    f. 8 in ×

    g. 14 in^2 ×

    h. 250 cm^3 ×

    Note: Think carefully about g and h. A picture may help.

36. Use Table 1.5 to assess whether or not the following statements are *reasonable*.
    a. Joe is 180 cm tall.

    b. I rode my bike to campus at a speed of 50 m/s.

    c. A skier reaches the bottom of the hill going 25 m/s.

d. I can throw a ball a distance of 2 km.

e. I can throw a ball at a speed of 50 km/h.

37. Justify the assertion that 1 m/s ≈ 2 mph by *exactly* converting 1 m/s to English units. By what percentage is this rough conversion in error?

38. How many significant figures does each of the following numbers have?

a. 6.21 _____   e. 0.0621 _____   i. 1.0621 _____

b. 62.1 _____   f. 0.620 _____   k. 6.21×10^3 _____

c. 6210 _____   g. 0.62 _____   j. 6.21×10^{-3} _____

d. 6210.0 _____   h. .62 _____   l. 6.21×10^3 _____

39. Compute the following numbers, applying the significant figure standards adopted for this text.

a. $33.3 \times 25.4 =$ _____   e. $2.345 \times 3.321 =$ _____

b. $33.3 - 25.4 =$ _____   f. $(4.32 \times 1.23) - 5.1 =$ _____

c. $33.3 \div 45.1 =$ _____   g. $33.3^2 =$ _____

d. $33.3 \times 45.1 =$ _____   h. $\sqrt{33.3} =$ _____

2 Kinematics in One Dimension

2.1 Uniform Motion

1. Sketch position-versus-time graphs (*x* versus *t* or *y* versus *t*) for the following motions. Include appropriate numerical scales along both axes. A small amount of computation may be necessary.

   a. A parachutist opens her parachute at an altitude of 1500 m. She then descends slowly to earth at a steady speed of 5 m/s. Start your graph as her parachute opens.

   b. Trucker Bob starts the day 120 miles west of Denver. He drives east for 3 hours at a steady 60 miles/hour before stopping for his coffee break. Let Denver be located at *x* = 0 mi and assume that the *x*-axis points to the east.

   c. Quarterback Bill throws the ball to the right at a speed of 15 m/s. It is intercepted 45 m away by Carlos, who is running to the left at 7.5 m/s. Carlos carries the ball 60 m to score. Let *x* = 0 m be the point where Bill throws the ball. Draw the graph for the *football*.

2. The figure shows a position-versus-time graph for the motion of objects A and B that are moving along the same axis.

    a. At the instant $t = 1$ s, is the speed of A greater than, less than, or equal to the speed of B? Explain.

    b. Do objects A and B ever have the *same* speed? If so, at what time or times? Explain.

3. Interpret the following position-versus-time graphs by writing a short "story" about what is happening. Your stories should make specific references to the *speeds* of the moving objects, which you can determine from the graphs. Assume that the motion takes place along a horizontal line.

    a. x (mi)

    b. x (mi)

4. Can the following be reasonably modeled as uniform motion? Answer Yes or No.

    a. An ice skater gliding across the ice.   _____

    b. A rocket being launched straight up.   _____

    c. A car braking for a stop sign.   _____

    d. A woman descending by parachute.   _____

2.2 Instantaneous Velocity

5. Draw both a position-versus-time graph *and* a velocity-versus-time graph for an object at rest at $x = 1$ m.

6. The figure shows the position-versus-time graphs for two objects, A and B, that are moving along the same axis.

   a. At the instant $t = 1$ s, is the speed of A greater than, less than, or equal to the speed of B? Explain.

   b. Do objects A and B ever have the *same* speed? If so, at what time or times? Explain.

7. Below are six position-versus-time graphs. For each, draw the corresponding velocity-versus-time graph directly below it. A vertical line drawn through both graphs should connect the velocity v_s at time t with the position s at the *same* time t. There are no numbers, but your graphs should correctly indicate the *relative* speeds.

   a.

   b.

c.

d.

e.

f.

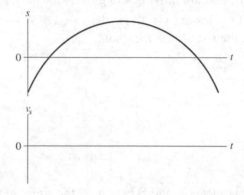

8. The figure shows a position-versus-time graph for a moving object. At which lettered point or points:

   a. Is the object *moving* the slowest? _____

   b. Is the object moving the fastest? _____

   c. Is the object at rest? _____

   d. Does the object have a constant
      nonzero velocity? _____

   e. Is the object moving to the left? _____

9. The figure shows a position-versus-time graph for a moving object. At which lettered point or points:

   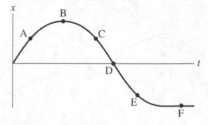

   a. Is the object moving the fastest? _____

   b. Is the object moving to the left? _____

   c. Is the object speeding up? _____

   d. Is the object slowing down? _____

   e. Is the object turning around? _____

10. For each of the following motions, draw
    - A motion diagam,
    - A position-versus-time graph, and
    - A velocity-versus-time graph.

    a. A car starts from rest, steadily speeds up to 40 mph in 15 s, moves at a constant speed for 30 s, then comes to a halt in 5 s.

    b. A rock is dropped from a bridge and steadily speeds up as it falls. It is moving at 30 m/s when it hits the ground 3 s later. Think carefully about the signs.

    c. A pitcher winds up and throws a baseball with a speed of 40 m/s. One-half second later the batter hits a line drive with a speed of 60 m/s. The ball is caught 1 s after it is hit. From where you are sitting, the batter is to the right of the pitcher. Draw your motion diagram and graph for the *horizontal* motion of the ball.

11. The figure shows six frames from the motion diagram of two moving cars, A and B.

    a. Draw both a position-versus-time graph and a velocity-versus-time graph. Show the motion of *both* cars on each graph. Label them A and B.

    b. Do the two cars ever have the same position at one instant of time?

       If so, in which frame number (or numbers)? _____

       Draw a vertical line through your graphs of part a to indicate this instant of time.

    c. Do the two cars ever have the same velocity at one instant of time?

       If so, between which two frames? _____

12. The figure shows six frames from the motion diagram of two moving cars, A and B.

    a. Draw both a position-versus-time graph and a velocity-versus-time graph. Show *both* cars on each graph. Label them A and B.

    b. Do the two cars ever have the same position at one instant of time?

       If so, in which frame number (or numbers)? _____

       Draw a vertical line through your graphs of part a to indicate this instant of time.

    c. Do the two cars ever have the same velocity at one instant of time?

       If so, between which two frames? _____

2.3 Finding Position from Velocity

13. Below are shown four velocity-versus-time graphs. For each:
    - Draw the corresponding position-versus-time graph.
    - Give a written description of the motion.

    Assume that the motion takes place along a horizontal line and that $x_0 = 0$.

a.

b.

c.

d.

14. The figure shows the velocity-versus-time graph for a moving object whose initial position is $x_0 = 20$ m. Find the object's position graphically, using the geometry of the graph, at the following times.

a. At $t = 3$ s.

b. At $t = 5$ s.

c. At $t = 7$ s.

d. You should have found a simple relationship between your answers to parts b and c. Can you explain this? What is the object doing?

2.4 Motion with Constant Acceleration

15. Give a specific example of motion for each of the following situations. Include both a description and a motion diagram.

    a. $a_x = 0$ but $v_x \neq 0$.

    b. $v_x = 0$ but $a_x \neq 0$.

    c. $v_x < 0$ and $a_x > 0$.

16. Can the following be reasonably modeled as constant acceleration? Answer Yes or No.

    a. A car braking for a stop sign. _____

    b. A rocket as it's running out of fuel. _____

    c. A skier gliding down a straight slope. _____

    d. A car turning a corner. _____

17. Below are three velocity-versus-time graphs. For each:
    • Draw the corresponding acceleration-versus-time graph.
    • Draw a motion diagram below the graphs.

    a.   b.   c.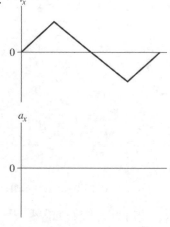

18. Below are three acceleration-versus-time graphs. For each, draw the corresponding velocity-versus-time graph. Assume that $v_{0x} = 0$.

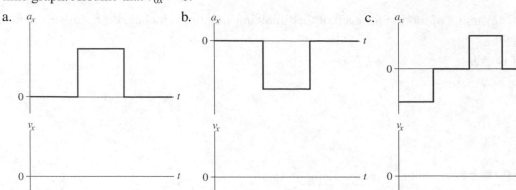

19. The figure below shows nine frames from the motion diagram of two cars. Both cars begin to accelerate, with constant acceleration, in frame 3.

a. Which car has the larger initial velocity? _____ The larger final velocity? _____

b. Which car has the larger acceleration after frame 3? How can you tell?

_____

c. Draw position, velocity, and acceleration graphs, showing the motion of both cars on each graph. (Label them A and B.) This is a total of three graphs with two curves on each.

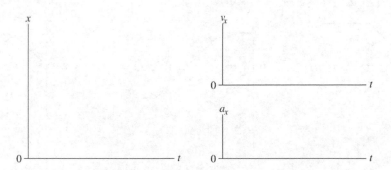

d. Do the cars ever have the same position at one instant of time? If so, in which frame? _____

e. Do the two cars ever have the same velocity at one instant of time? _____

If so, identify the *two* frames between which this velocity occurs. _____

Identify this instant on your graphs by drawing a vertical line through the graphs.

2.5 Free Fall

20. A ball is thrown straight up into the air. At each of the following instants, is the magnitude of the ball's acceleration greater than g, equal to g, less than g, or zero?

    a. Just after leaving your hand? _____

    b. At the very top (maximum height)? _____

    c. Just before hitting the ground? _____

21. A rock is *thrown* (not dropped) straight down from a bridge into the river below.

    a. Immediately *after* being released, is the magnitude of the rock's acceleration greater than g, less than g, or equal to g? Explain.

    b. Immediately before hitting the water, is the magnitude of the rock's acceleration greater than g, less than g, or equal to g? Explain.

22. A model rocket is launched straight up with constant acceleration a. It runs out of fuel at time t.
PSS Suppose you need to determine the maximum height reached by the rocket. We'll assume that air
2.1 resistance is negligible.

a. Is the rocket at maximum height the instant it runs out of fuel? _____

b. Is there anything other than gravity acting on the rocket after it runs out of fuel? _____

c. What is the name of motion under the influence of only gravity? _____

d. Draw a pictorial representation for this problem. You should have three identified points in the motion: launch, out of fuel, maximum height. Call these points 1, 2, and 3.

- Using subscripts, define 11 quantities: y, v_y, and t at each of the three points, plus acceleration a_1 connecting points 1 and 2 and acceleration a_2 connecting points 2 and 3.
- Identify 7 of these quantities as Knowns, either 0 or given symbolically in terms of a, t, and g. Be careful with signs!
- Identify which one of the 4 unknown quantities you're trying to find.

e. This is a two-part problem. Write two kinematic equations for the first part of the motion to determine—again symbolically—the two unknown quantities at point 2.

f. Now write a kinematic equation for the second half of the motion that will allow you to find the desired unknown that will answer the question. Your equation should not contain the fourth unknown quantity. Just write the equation; don't solve it yet.

g. Now, substitute what you learned in part e into your equation of part f, do the algebra to solve for the unknown, and simplify the result as much as possible.

2.6 Motion on an Inclined Plane

23. A ball released from rest on an inclined plane accelerates down the plane at 2 m/s². Complete the table below showing the ball's velocities at the times indicated. Do *not* use a calculator for this; this is a reasoning question, not a calculation problem.

Time (s)	Velocity (m/s)
0	0
1	_____
2	_____
3	_____
4	_____
5	_____

24. A bowling ball rolls along a level surface, then up a 30° slope, and finally exits onto another level surface at a much slower speed.

a. Draw position-, velocity-, and acceleration-versus-time graphs for the ball.

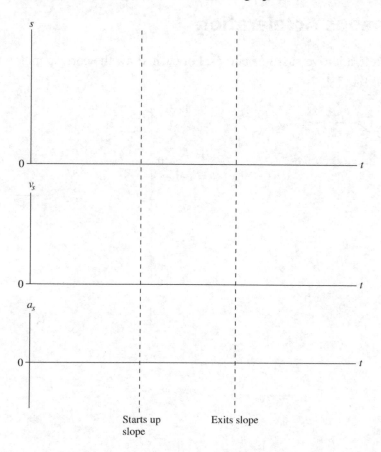

b. Suppose that the ball's initial speed is 5.0 m/s and its final speed is 1.0 m/s. Draw a pictorial representation that you would use to determine the height h of the slope. Establish a coordinate system, define all symbols, list known information, and identify desired unknowns.

Note: Don't actually solve the problem. Just draw the complete pictorial representation that you would use as a first step in solving the problem.

2.7 Instantaneous Acceleration

25. Below are two acceleration-versus-time curves. For each, draw the corresponding velocity-versus-time curve. Assume that $v_{0x} = 0$.

a.

b.

3 Vectors and Coordinate Systems

3.1 Vectors

3.2 Properties of Vectors

Exercises 1–3: Draw and label the vector sum $\vec{A} + \vec{B}$.

1.

2.

3.

4. Use a figure and the properties of vector addition to show that vector addition is associative. That is, show that

$$(\vec{A} + \vec{B}) + \vec{C} = \vec{A} + (\vec{B} + \vec{C})$$

Exercises 5–7: Draw and label the vector difference $\vec{A} - \vec{B}$.

5.

6.

7.

8. Draw and label the vector $2\vec{A}$ and the vector $\frac{1}{2}\vec{A}$.

9. Given vectors $\vec{A}$ and $\vec{B}$ below, find the vector $\vec{C} = 2\vec{A} - 3\vec{B}$.

3.3 Coordinate Systems and Vector Components

Exercises 10–12: Draw and label the *x*- and *y*-component vectors of the vector shown.

10.   11.   12.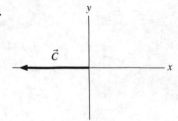

Exercises 13–15: Determine the numerical values of the *x*- and *y*-components of each vector.

13.   14.   15.

$A_x =$ _____ $B_x =$ _____ $C_x =$ _____

$A_y =$ _____ $B_y =$ _____ $C_y =$ _____

Exercises 16–18: Draw and label the vector with these components. Then determine the magnitude of the vector.

16. $A_x = 3, A_y = -2$

17. $B_x = -2, B_y = 2$

18. $C_x = 0, C_y = -2$

$A =$ _____

$B =$ _____

$C =$ _____

3.4 Vector Algebra

Exercises 19–21: Draw and label the vectors on the axes.

19. $\vec{A} = -\hat{\imath} + 2\hat{\jmath}$

20. $\vec{B} = -2\hat{\jmath}$

21. $\vec{C} = 3\hat{\imath} - 2\hat{\jmath}$

Exercises 22–24: Write the vector in component form (e.g., $3\hat{\imath} + 2\hat{\jmath}$).

22.

$\vec{A} =$ _____

23.

$\vec{B} =$ _____

24.

$\vec{C} =$ _____

25. What is the vector sum $\vec{D} = \vec{A} + \vec{B} + \vec{C}$ of the three vectors defined in Exercises 22–24? Write your answer in *component* form.

Exercises 26–28: For each vector:
- Draw the vector on the axes provided.
- Draw and label an angle θ to describe the direction of the vector.
- Find the magnitude and the angle of the vector.

26. $\vec{A} = 2\hat{\imath} + 2\hat{\jmath}$

27. $\vec{B} = -2\hat{\imath} + 2\hat{\jmath}$

28. $\vec{C} = 3\hat{\imath} + \hat{\jmath}$

$A =$ _____

$\theta =$ _____

$B =$ _____

$\theta =$ _____

$C =$ _____

$\theta =$ _____

Exercises 29–31: Define vector $\vec{A} = (5, 30° \text{ above the horizontal})$. Determine the components A_x and A_y in the three coordinate systems shown below. Show your work below the figure.

29.

30.

31.

$A_x =$ _____

$A_y =$ _____

$A_x =$ _____

$A_y =$ _____

$A_x =$ _____

$A_y =$ _____

4 Kinematics in Two Dimensions

4.1 Motion in Two Dimensions

Exercises 1–2: The figures below show an object's position in three successive frames of a video. The object is moving in the direction $0 \rightarrow 1 \rightarrow 2$. For each diagram:
- Draw and label the initial and final velocity vectors $\vec{v}_0$ and $\vec{v}_1$ Use **black**.
- Use the steps of Tactics Box 4.1 to find the change in velocity $\Delta\vec{v}$.
- Draw and label $\vec{a}$ next to dot 1 on the motion diagram. Use **red**.
- Determine whether the object is speeding up, slowing down, or moving at a constant speed. Write your answer beside the diagram.

3. The figure shows a ramp and a ball that rolls along the ramp. Draw vector arrows on the figure to show the ball's acceleration at each of the lettered points A to E (or write $\vec{a} = \vec{0}$, if appropriate).
   Hint: At each point is the ball changing speed, changing direction, or both? Be especially careful at point D.

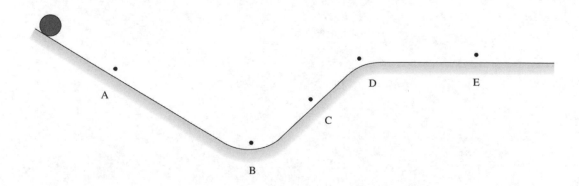

4. Complete the motion diagram for this trajectory, showing velocity and acceleration vectors.

Start ●
●
●
●
●
●
●

Exercises 5–6: Draw a complete motion diagram for each of the following.
- Draw and label the velocity vectors $\vec{v}$. Use **black**.
- Draw and label the acceleration vectors $\vec{a}$. Use **red**.

5. A cannon ball is fired from a Civil War cannon up onto a high cliff. Show the cannon ball's motion from the instant it leaves the cannon until a microsecond before it hits the ground.

6. A plane flying north at 300 mph turns slowly to the west without changing speed, then continues to fly west. Draw the motion diagram from a viewpoint above the plane.

7. A particle moving in the xy-plane has the x-versus-t graph and the y-versus-t graphs shown below. Use the grid to draw a y-versus-x graph of the trajectory.

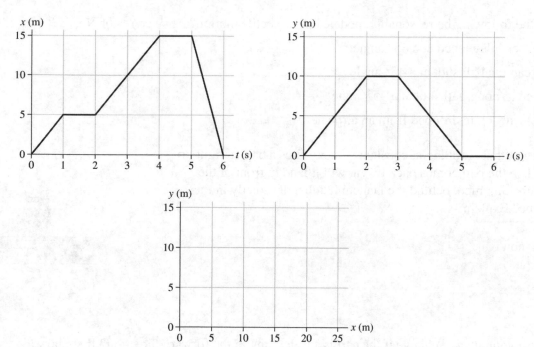

8. The trajectory of a particle is shown below. The particle's position is indicated with dots at 1-second intervals. The particle moves between each pair of dots at constant speed. Draw x-versus-t and y-versus-t graphs for the particle.

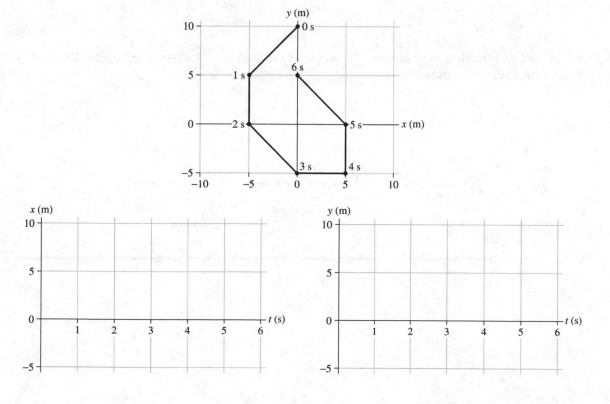

4.2 Projectile Motion

9. Can the following be reasonably modeled as projectile motion? Answer Yes or No.

   a. A rocket launched at a 45° angle. _____

   b. A cannon ball shot at a 45° angle. _____

   c. A ping-pong ball shot at a 45° angle. _____

   d. A heavy crate dropped from an airplane. _____

10. a. A cart that is rolling at constant velocity fires a ball straight up. When the ball comes back down, will it land in front of the launching tube, behind the launching tube, or directly in the tube? Explain.

    b. Will your answer change if the cart is accelerating in the forward direction? If so, how?

11. Rank in order, from shortest to longest, the amount of time it takes each of these projectiles to hit the ground. Ignore air resistance. (Some may be simultaneous.)

    Order:

    Explanation:

4.3 Relative Motion

12. Anita is running to the right at 5 m/s. Balls 1 and 2 are thrown toward her at 10 m/s by friends standing on the ground. According to Anita, which ball is moving faster? Or are both speeds the same? Explain.

13. Anita is running to the right at 5 m/s. Balls 1 and 2 are thrown toward her by friends standing on the ground. According to Anita, both balls are approaching her at 10 m/s. Which ball was thrown at a faster speed? Or were they thrown with the same speed? Explain.

14. Ryan, Samantha, and Tomas are driving their convertibles at a steady speed. At the same instant, they each see a jet plane with an instantaneous velocity of 200 m/s and an acceleration of 5 m/s² relative to the ground. Rank in order, from largest to smallest, the jet's *speed* v_R, v_S, and v_T according to Ryan, Samantha, and Tomas. Explain.

15. An electromagnet on the ceiling of an airplane holds a steel ball. When a button is pushed, the magnet releases the ball. The experiment is first done while the plane is parked on the ground, and the point where the ball hits the floor is marked with an X. Then the experiment is repeated while the plane is flying level at a steady 500 mph. Does the ball land slightly in front of the X (toward the nose of the plane), on the X, or slightly behind the X (toward the tail of the plane)? Explain.

16. Zack is driving past his house. He wants to toss his physics book out the window and have it land in his driveway. If he lets go of the book exactly as he passes the end of the driveway, should he direct his throw outward and toward the front of the car (throw 1), straight outward (throw 2), or outward and toward the back of the car (throw 3)? Explain. (Ignore air resistance.)

17. Yvette and Zack are driving down the freeway side by side with their windows rolled down. Zack wants to toss his physics book out the window and have it land in Yvette's front seat. Should he direct his throw outward and toward the front of the car (throw 1), straight outward (throw 2), or outward and toward the back of the car (throw 3)? Explain. (Ignore air resistance.)

4.4 Uniform Circular Motion

4.5 Centripetal Acceleration

18. a. The crankshaft in your car rotates at 3000 rpm. What is the frequency in revolutions per second?

b. A record turntable rotates at 33.3 rpm. What is the period in seconds?

19. The figure shows three points on a steadily rotating wheel.

a. Draw the velocity vectors at each of the three points

b. Rank in order, from largest to smallest, the angular velocities ω_1, ω_2, and ω_3 of these points.

Order:
Explanation:

c. Rank in order, from largest to smallest, the speeds v_1, v_2, and v_3 of these points.

Order:
Explanation:

20. Can the following be reasonably modeled as uniform circular motion? Answer Yes or No.

a. A horse on a moving merry-go-round. _____

b. A rock stuck in the tire of a coasting bicycle. _____

c. A rock stuck in the tire of a braking car. _____

d. A point on a wheel rolling down a ramp. _____

21. Below are two angular position-versus-time graphs. For each, draw the corresponding angular velocity-versus-time graph directly below it.

a.

b.
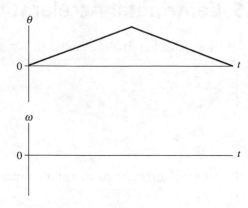

22. Below are two angular velocity-versus-time graphs. For each, draw the corresponding angular position-versus-time graph directly below it. Assume $\theta_0 = 0$ rad.

a.

b.

23. A particle in circular motion rotates clockwise at 4 rad/s for 2 s, then counterclockwise at 2 rad/s for 4 s. The time required to change direction is negligible. Graph the angular velocity and the angular position, assuming $\theta_0 = 0$ rad.

24. A particle in uniform circular motion has $a = 8$ m/s^2. What is a if

a. The radius is doubled without changing the angular velocity? _____

b. The radius is doubled without changing the particle's speed? _____

c. The angular velocity is doubled without changing the circle's radius? _____

4.6 Nonuniform Circular Motion

25. The following figures show a rotating wheel. Determine the signs (+ or −) of ω and α.

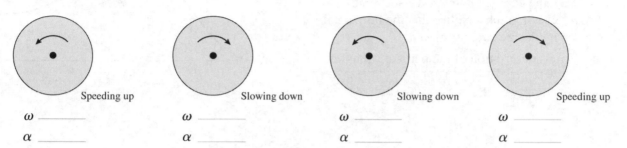

| Speeding up | Slowing down | Slowing down | Speeding up |

ω _____ ω _____ ω _____ ω _____

α _____ α _____ α _____ α _____

26. The figures below show the radial acceleration vector $\vec{a}_r$ at four successive points on the trajectory of a particle moving in a counterclockwise circle.
    a. For each, draw the tangential acceleration vector $\vec{a}_t$ at points 2 and 3 or, if appropriate, write $\vec{a}_t = \vec{0}$.
    b. Determine if the particle's angular acceleration α is positive (+), negative (−), or zero (0).

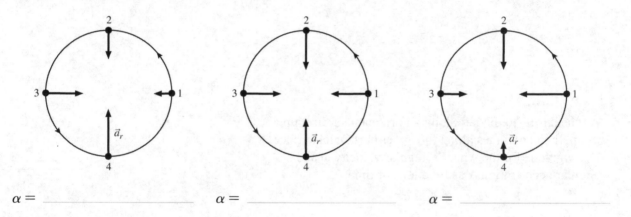

$\alpha = $ _____ $\alpha = $ _____ $\alpha = $ _____

27. A pendulum swings from its end point on the left (point 1) to its end point on the right (point 5). At each of the labeled points:

    a. Use a **black** pen or pencil to draw and label the vectors $\vec{a}_r$ and $\vec{a}_t$ at each point. Make sure the length indicates the relative size of the vector.
    b. Use a **red** pen or pencil to draw and label the total acceleration vector $\vec{a}$.

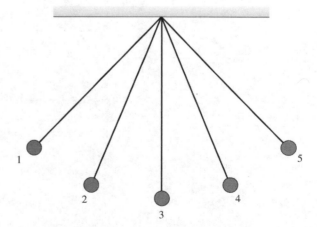

28. The figure shows the θ-versus-*t* graph for a particle moving in a circle. The curves are all sections of parabolas.

    a. Draw the corresponding ω-versus-*t* and α-versus-*t* graphs. Notice that the horizontal tick marks are equally spaced.

    b. Write a description of the particle's motion.

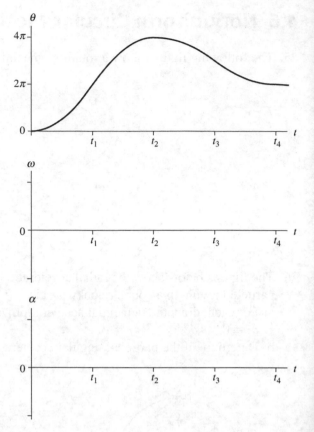

29. A wheel rolls to the left along a horizontal surface, up a ramp, then continues along the upper horizontal surface. Draw graphs for the wheel's angular velocity ω and angular acceleration α as functions of time.

5 Force and Motion

5.1 Force

1. Two or more forces are shown on the objects below. Draw and label the net force $\vec{F}_{net}$.

2. Two or more forces are shown on the objects below. Draw and label the net force $\vec{F}_{net}$.

5.2 A Short Catalog of Forces

5.3 Identifying Forces

Exercises 3–8: Follow the six-step procedure of Tactics Box 5.2 to identify and name all the forces acting on the object.

3. An elevator suspended by a cable is descending at constant velocity.

4. A car on a *very* slippery icy road is sliding headfirst into a snowbank, where it gently comes to rest with no one injured. (Question: What does "*very* slippery" imply?)

5. A compressed spring is pushing a block across a rough horizontal table.

6. A brick is falling from the roof of a three-story building.

7. Blocks A and B are connected by a string passing over a pulley. Block B is falling and dragging block A across a frictionless table. Analyze block A.

8. A rocket is launched at a 30° angle. Air resistance is not negligible.

5.4 What Do Forces Do?

9. The figure shows an acceleration-versus-force graph for an object of mass m. Data have been plotted as individual points, and a line has been drawn through the points.

   Draw and label, directly on the figure, the acceleration-versus-force graphs for objects of mass

   a. $2m$ b. $0.5m$

   Use triangles ▲ to show four points for the object of mass $2m$, then draw a line through the points. Use squares ■ for the object of mass $0.5m$.

10. A constant force applied to object A causes A to accelerate at 5 m/s². The same force applied to object B causes an acceleration of 3 m/s². Applied to object C, it causes an acceleration of 8 m/s².

    a. Which object has the largest mass? _____

    b. Which object has the smallest mass? _____

    c. What is the ratio of mass A to mass B? $(m_A/m_B) =$ _____

11. A constant force applied to an object causes the object to accelerate at 10 m/s². What will the acceleration of this object be if

    a. The force is doubled? _____   b. The mass is doubled? _____

    c. The force is doubled *and* the mass is doubled? _____

    d. The force is doubled *and* the mass is halved? _____

12. A constant force applied to an object causes the object to accelerate at 8 m/s². What will the acceleration of this object be if

    a. The force is halved? _____   b. The mass is halved? _____

    c. The force is halved *and* the mass is halved? _____

    d. The force is halved *and* the mass is doubled? _____

5.5 Newton's Second Law

13. Forces are shown on two objects. For each:

    a. Draw and label the net force vector. Do this right on the figure.
    b. Below the figure, draw and label the object's acceleration vector.

14. Forces are shown on two objects. For each:
    a. Draw and label the net force vector. Do this right on the figure.
    b. Below the figure, draw and label the object's acceleration vector.

15. In the figures below, one force is missing. Use the given direction of acceleration to determine the missing force and draw it on the object. Do all work directly on the figure.

16. Below are two motion diagrams for a particle. Draw and label the net force vector at point 2.

17. Below are two motion diagrams for a particle. Draw and label the net force vector at point 2.

5.6 Newton's First Law

18. If an object is at rest, can you conclude that there are no forces acting on it? Explain.

19. If a force is exerted on an object, is it possible for that object to be moving with constant velocity? Explain.

20. A hollow tube forms three-quarters of a circle. It is lying flat on a table. A ball is shot through the tube at high speed. As the ball emerges from the other end, does it follow path A, path B, or path C? Explain your reasoning.

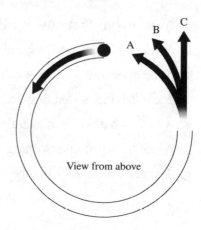

View from above

21. Which, if either, of the objects shown below is in equilibrium? Explain your reasoning.

22. Two forces are shown on the objects below. Add a third force $\vec{F}_3$ that will cause the object to be in equilibrium.

23. Are the following inertial reference frames? Answer Yes or No.

    a. A car driving at steady speed on a straight and level road. _____

    b. A car driving at steady speed up a 10° incline. _____

    c. A car speeding up after leaving a stop sign. _____

    d. A car driving at steady speed around a curve. _____

    e. A hot air balloon rising straight up at steady speed. _____

    f. A skydiver just after leaping out of a plane. _____

    g. A space station orbiting the earth.

5.7 Free-Body Diagrams

Exercises 24–29:

- Draw a picture and identify the forces, then
- Draw a complete free-body diagram for the object, following each of the steps given in Tactics Box 5.3. Be sure to think carefully about the direction of $\vec{F}_{net}$.

Note: Draw individual force vectors with a **black** or **blue** pencil or pen. Draw the *net* force vector $\vec{F}_{net}$ with a **red** pencil or pen.

24. A heavy crate is being lowered straight down at a constant speed by a steel cable.

25. A boy is pushing a box across the floor at a steadily increasing speed. Let the box be "the system" for analysis.

26. A bicycle is speeding up down a hill. Friction is negligible, but air resistance is not.

27. You've slammed on your car brakes while going down a hill. You're skidding to a halt.

28. You are going to toss a rock *straight up* into the air by placing it on the palm of your hand (you're not gripping it), then pushing your hand up very rapidly. You may want to toss an object into the air this way to help you think about the situation. The rock is "the system" of interest.

 a. As you hold the rock at rest on your palm, before moving your hand.

 b. As your hand is moving up but before the rock leaves your hand.

 c. One-tenth of a second after the rock leaves your hand.

 d. After the rock has reached its highest point and is now falling straight down.

29. Block B has just been released and is beginning to fall. The table has friction. Analyze block A.

Dynamics I: Motion Along a Line

6.1 The Equilibrium Model

1. The vectors below show five forces that can be applied individually or in combinations to an object. Which forces or combinations of forces will cause the object to be in equilibrium?

2. The free-body diagrams show a force or forces acting on an object. Draw and label one more force (one that is appropriate to the situation) that will cause the object to be in equilibrium.

3. If you know all of the forces acting on a moving object, can you tell in which direction the object is moving? If the answer is Yes, explain how. If the answer is No, give an example.

6.2 Using Newton's Second Law

4. a. An elevator travels *upward* at a constant speed. The elevator hangs by a single cable. Friction and air resistance are negligible. Is the tension in the cable greater than, less than, or equal to the weight of the elevator? Explain. Your explanation should include both a free-body diagram and reference to appropriate physical principles.

   b. The elevator travels *downward* and is slowing down. Is the tension in the cable greater than, less than, or equal to the weight of the elevator? Explain.

Exercises 5–6: The figures show free-body diagrams for an object of mass *m*. Write the *x*- and *y*-components of Newton's second law. Write your equations in terms of the *magnitudes* of the forces F_1, F_2, ... and any *angles* defined in the diagram. One equation is shown to illustrate the procedure.

5.

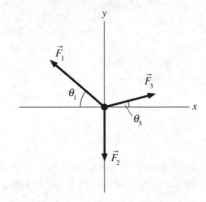

$ma_x = $ _____

$ma_y = F_1 - F_2$

$ma_x = $ _____

$ma_y = $ _____

6.

$$ma_x = F_3 \cos \theta_3 - F_4$$

$$ma_y = \underline{\hspace{4cm}}$$

$$ma_x = \underline{\hspace{4cm}}$$

$$ma_y = \underline{\hspace{4cm}}$$

Exercises 7–9: Two or more forces, shown on a free-body diagram, are exerted on a 2 kg object. The units of the grid are newtons. For each:

• Draw a vector arrow *on the grid,* starting at the origin, to show the net force $\vec{F}_{net}$.

• In the space to the right, determine the numerical values of the components a_x and a_y.

7.

$$a_x = \underline{\hspace{4cm}}$$

$$a_y = \underline{\hspace{4cm}}$$

8.

$$a_x = \underline{\hspace{4cm}}$$

$$a_y = \underline{\hspace{4cm}}$$

9.

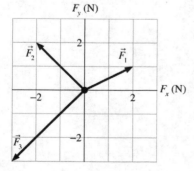

$$a_x = \underline{\hspace{4cm}}$$

$$a_y = \underline{\hspace{4cm}}$$

Exercises 10–12: Three forces $\vec{F}_1$, $\vec{F}_2$, and $\vec{F}_3$ cause a 1 kg object to accelerate with the acceleration given. Two of the forces are shown on the free-body diagrams below, but the third is missing. For each, draw and label *on the grid* the missing third force vector.

10. $\vec{a} = 2\hat{\imath}$ m/s^2

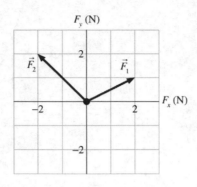

11. $\vec{a} = -3\hat{\jmath}$ m/s^2

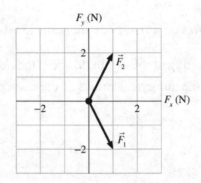

12. The object moves with constant velocity.

13. Three arrows are shot horizontally. They have left the bow and are traveling parallel to the ground. Air resistance is negligible. Rank in order, from largest to smallest, the magnitudes of the *horizontal* forces $\vec{F}_1$, $\vec{F}_2$, and $\vec{F}_3$ acting on the arrows. Some may be equal. Give your answer in the form A > B = C > D.

Order:

Explanation:

6.3 Mass, Weight, and Gravity

14. An astronaut takes his bathroom scales to the moon and then stands on them. Is the reading of the scales his weight? Explain.

15. Suppose you attempt to pour out 100 g of salt, using a pan balance for measurement, while in an elevator that is accelerating upward. Will the quantity of salt be too much, too little, or the correct amount? Explain.

16. An astronaut orbiting the earth is handed two balls that are identical in outward appearance. However, one is hollow while the other is filled with lead. How might the astronaut determine which is which? Cutting them open is not allowed.

17. The terms "vertical" and "horizontal" are frequently used in physics. Give operational definitions for these two terms. An operational definition defines a term by how it is measured or determined. Your definition should apply equally well in a laboratory or on a steep mountainside.

18. Suppose you stand on a spring scale in six identical elevators. Each elevator moves as shown below. Let the reading of the scale in elevator n be S_n. Rank in order, from largest to smallest, the six scale readings S_1 to S_6. Some may be equal. Give your answer in the form A > B = C > D.

Order:

Explanation:

6.4 Friction

19. A block pushed along the floor with velocity $\vec{v}_0$ slides a distance d after the pushing force is removed.

    a. If the mass of the block is doubled but the initial velocity is not changed, what is the distance the block slides before stopping? Explain.

    b. If the initial velocity of the block is doubled to $2\vec{v}_0$ but the mass is not changed, what is the distance the block slides before stopping? Explain.

20. Suppose you press a book against the wall with your hand. The book is not moving.

    a. Identify the forces on the book and draw a free-body diagram.

    b. Now suppose you decrease your push, but not enough for the book to slip. What happens to each of the following forces? Do they increase in magnitude, decrease, or not change?

    $\vec{F}_{push}$ _____

    $\vec{F}_G$ _____

    $\vec{n}$ _____

    $\vec{f}_s$ _____

    $\vec{f}_{s\ max}$ _____

21. Consider a box in the back of a pickup truck.

    a. If the truck accelerates slowly, the box moves with the truck without slipping. What force or forces act on the box to accelerate it? In what direction do those forces point?

    b. Draw a free-body diagram of the box.

    c. What happens to the box if the truck accelerates too rapidly? Explain why this happens, basing your explanation on physical models and the principles described in this chapter.

22. A car is parked on a road that slopes upward at angle θ. The magnitude of the normal force of the road on the car is $mg\cos\theta$. Is the magnitude of the static friction force on the car less than, equal to, or greater than $\mu_s mg\cos\theta$? Explain.

23. A small airplane of mass m must take off from a primitive jungle airstrip that slopes upward at a slight
PSS angle θ. When the pilot pulls back on the throttle, the plane's engines exert a constant forward force $\vec{F}_{thrust}$.
6.1 Rolling friction is not negligible on the dirt airstrip, and the coefficient of rolling resistance is μ_r. If the
plane's take-off speed is v_{off}, what minimum length must the airstrip have for the plane to get airborne?

a. Assume the plane takes off uphill to the right. Begin with a pictorial representation, as was
described in Tactics Box 1.5. Establish a coordinate system with a tilted x-axis; show the plane at the
beginning and end of the motion; define symbols for position, velocity, and time at these two points
(six symbols all together); list known information; and state what you wish to find. $\vec{F}_{thrust}$, m, θ, μ_r,
and v_{off} are presumed known, although we have only symbols for them rather than numerical values,
and three other quantities are zero.

b. Next, draw a force-identification diagram. Beside it, draw a free-body diagram. Your free-body
diagram should use the same coordinate system you established in part a, and it should have 4 forces
shown on it.

c. Write Newton's second law as two equations, one for the net force in the x-direction and one for the
net force in the y-direction. Be careful finding the components of $\vec{F}_G$ (see Figure 6.2), and pay close
attention to signs. Remember that symbols such as F_G or f_r or represent the *magnitudes* of vectors;
you have to supply appropriate signs to indicate which way the vectors point. The right side of these
equations have a_x and a_y. The motion is entirely along the x-axis, so what do you know about a_y?
Use this information as you write the y-equation.

d. Now write the equation that characterizes the friction force on a rolling tire.

e. Combine your friction equation with the *y*-equation of Newton's second law to find an expression for the magnitude of the friction force.

f. Finally, substitute your answer to part e into the *x*-equation of Newton's second law, and then solve for a_x, the *x*-component of acceleration. Use $F_G = mg$ if you've not already done so.

g. With friction present, should the *magnitude* of the acceleration be larger or smaller than the acceleration of taking off on a frictionless runway? _____

h. Does your expression for acceleration agree with your answer to part g? _____
Explain how you can tell. If it doesn't, recheck your work.

i. The force analysis is done, but you still have to do the kinematics. This is a situation where we know about velocities, distance, and acceleration but nothing about the time involved. That should suggest the appropriate kinematics equation. Use your acceleration from part f in that kinematics equation, and solve for the unknown quantity you're seeking.

You've found a symbolic answer to the problem, one that you could now evaluate for a range of values of F_{thrust} or θ without having to go through the entire solution each time.

6.5 Drag

24. Three objects move through the air as shown. Rank in order, from largest to smallest, the three drag forces $F_{drag\ 1}$, $F_{drag\ 2}$, and $F_{drag\ 3}$. Some may be equal. Give your answer in the form $A > B = C > D$.

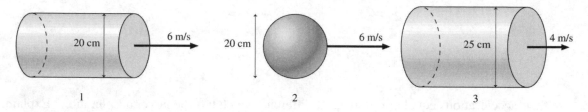

Order:

Explanation:

25. Five balls move through the air as shown. All five have the same size and shape. Rank in order, from largest to smallest, the magnitude of their accelerations a_1 to a_5. Some may be equal. Give your answer in the form $A > B = C > D$.

Order:

Explanation:

26. A 1 kg wood ball and a 10 kg lead ball have identical shapes and sizes. They are dropped simultaneously from a tall tower.

   a. To begin, assume that air resistance is negligible. As the balls fall, are the forces on them equal in magnitude or different? If different, which has the larger force? Explain.

   b. Are their accelerations equal or different? If different, which has the larger acceleration? Explain.

   c. Which ball hits the ground first? Or do they hit simultaneously? Explain.

   d. If air resistance is present, each ball will experience the *same* drag force because both have the same shape. Draw free-body diagrams for the two balls as they fall in the presence of air resistance. Make sure that your vectors all have the correct *relative* lengths.

   e. When air resistance is included, are the accelerations of the balls equal or different? If not, which has the larger acceleration? Explain, using your free-body diagrams and Newton's laws.

   f. Which ball now hits the ground first? Or do they hit simultaneously? Explain.

7 Newton's Third Law

7.1 Interacting Objects

7.2 Analyzing Interacting Objects

Exercises 1–7: Follow steps 1–3 of Tactics Box 7.1 to draw interaction diagrams describing the following situations. Your diagrams should be similar to Figures 7.5 and 7.9.

1. A bat hits a ball.

2. A massless string pulls a box across the floor. Friction is not negligible.

3. A boy pulls a wagon by a rope attached to the front of the wagon. The rope is not massless, and rolling friction is not negligible.

4. A skateboarder is pushing on the ground to speed up. Treat the person and the skateboard as separate objects.

5. The bottom block is pulled by a massless string. Friction is not negligible. Treat the two blocks as separate objects.

6. A crate in the back of a truck does not slip as the truck accelerates forward. Treat the crate and the truck as separate objects.

7. The bottom block is pulled by a massless string. Friction is not negligible. Treat the pulley as a separate object.

7.3 Newton's Third Law

8. Block A is pushed across a horizontal surface at a *constant* speed by a hand that exerts force $\vec{F}_{\text{H on A}}$. The surface has friction.

Hand

A

a. Draw two free-body diagrams, one for the hand and the other for the block. On these diagrams:

   • Show only the *horizontal* forces, such as was done in Figure 7.13 of the text.
   • Label force vectors, using the form $\vec{F}_{\text{C on D}}$.
   • Connect action/reaction pairs with dotted lines.
   • On the hand diagram show both $\vec{F}_{\text{A on H}}$ and $\vec{F}_{\text{arm on H}}$.
   • Make sure vector lengths correctly portray the relative magnitudes of the forces.

b. Rank in order, from largest to smallest, the magnitudes of *all* of the horizontal forces you showed in part a. For example, if $F_{\text{C on D}}$ is the largest of three forces while $F_{\text{D on C}}$ and $F_{\text{D on E}}$ are smaller but equal, you can record this as $F_{\text{C on D}} > F_{\text{D on C}} = F_{\text{D on E}}$.

Order:

Explanation:

c. Repeat both part a and part b for the case that the block is *speeding up*.

9. A second block B is placed in front of Block A of question 8. B is more massive than A: $m_B > m_A$. The blocks are speeding up.

   a. Consider a *frictionless* surface. Draw *separate* free-body diagrams for A, B, and the hand H. Show only the horizontal forces. Label forces in the form $\vec{F}_{C \text{ on } D}$. Use dashed lines to connect action/reaction pairs.

   b. By applying the second law to each block and the third law to each action/reaction pair, rank in order *all* of the horizontal forces, from largest to smallest.

   Order:
   Explanation:

   c. Repeat parts a and b if the surface has friction. Assume that A and B have the same coefficient of kinetic friction.

10. Blocks A and B are held on the palm of your outstretched hand as you lift them straight up at *constant speed.* Assume $m_B > m_A$ and that $m_{hand} = 0$.

    a. Draw *separate* free-body diagrams for A, B, and your hand H.

       • Show *all* vertical forces, including the gravitational forces on the blocks. Also include the force $\vec{F}_{arm\ on\ H}$.
       • Make sure vector lengths indicate the relative sizes of the forces.
       • Label forces in the form $\vec{F}_{C\ on\ D}$.
       • Connect action/reaction pairs with dashed lines.

    b. Rank in order, from largest to smallest, all of the vertical forces. Explain your reasoning.

11. A mosquito collides head-on with a car traveling 60 mph.

    a. How do you think the size of the force that the car exerts on the mosquito compares to the size of the force that the mosquito exerts on the car?

    b. Draw *separate* free-body diagrams of the car and the mosquito at the moment of collision, showing only the horizontal forces. Label forces in the form $\vec{F}_{C\ on\ D}$. Connect action/reaction pairs with dotted lines.

Exercises 12–16: Write the acceleration constraint in terms of *components*. For example, write $(a_1)_x = (a_2)_y$, if that is the appropriate answer, rather than $\vec{a}_1 = \vec{a}_2$.

12.

Constraint: _____

13.

Constraint: _____

14.

Constraint: _____

15.

Constraint: _____

16.

Constraint: _____

7.4 Ropes and Pulleys

Exercises 17–22: Determine the reading of the spring scale.
- All the masses are at rest.
- The strings and pulleys are massless, and the pulleys are frictionless.
- The spring scale reads in kg.

17.

Scale = _____

18.

Scale = _____

19.

Scale = _____

20.

Scale = _____

21.

Scale = _____

22.

Scale = _____

7.5 Examples of Interacting-Objects Problems

23. Blocks A and B, with $m_B > m_A$, are connected by a string. A hand pushing on the back of A accelerates them along a frictionless surface. The string (S) is massless.

a. Draw separate free-body diagrams for A, S, and B, showing only horizontal forces. Be sure vector lengths indicate the relative size of the force. Connect any action/reaction pairs with dotted lines.

b. Rank in order, from largest to smallest, all of the horizontal forces. Explain.

c. Repeat parts a and b if the string has mass.

d. You might expect to find $F_{S \text{ on } B} > F_{H \text{ on } A}$ because $m_B > m_A$. Did you? Explain why $F_{S \text{ on } B} > F_{H \text{ on } A}$ is or is not a correct statement.

24. Blocks A and B are connected by a massless string over a massless, frictionless pulley. The blocks have just this instant been released from rest.

    a. Will the blocks accelerate? If so, in which directions?

    Pulley

    A 2 kg

    b. Draw a separate free-body diagram for each block. Be sure vector lengths indicate the relative size of the force. Connect any action/reaction pairs or "as if" pairs with dashed lines.

    B 1 kg

    c. Rank in order, from largest to smallest, all of the vertical forces. Explain.

    d. Compare the magnitude of the *net* force on A with the *net* force on B. Are they equal, or is one larger than the other? Explain.

    e. Consider the block that falls. Is the magnitude of its acceleration less than, greater than, or equal to *g*? Explain.

25. In case a, block A is accelerated across a frictionless table by a hanging 10 N weight (1.02 kg). In case b, the same block is accelerated by a steady 10 N tension in the string.

Is block A's acceleration in case b greater than, less than, or equal to its acceleration in case a? Explain.

Exercises 26–27: Draw separate free-body diagrams for blocks A and B. Connect any action/reaction pairs (or forces that act *as if* they are action/reaction pairs) together with dashed lines.

26.

27.

Dynamics II: Motion in a Plane

8.1 Dynamics in Two Dimensions

1. An ice hockey puck is pushed across frictionless ice in the direction shown. The puck receives a sharp, very short-duration kick toward the right as it crosses line 2. It receives a second kick, of equal strength and duration but toward the left, as it crosses line 3. Sketch the puck's trajectory from line 1 until it crosses line 4.

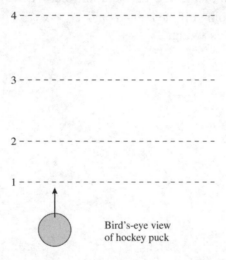

Bird's-eye view
of hockey puck

2. A rocket motor is taped to an ice hockey puck, oriented so that the thrust is to the left. The puck is given a push across frictionless ice in the direction shown. The rocket will be turned on by remote control as the puck crosses line 2, then turned off as it crosses line 3. Sketch the puck's trajectory from line 1 until it crosses line 4.

Bird's-eye view
of hockey puck

3. An ice hockey puck is sliding from west to east across frictionless ice. When the puck reaches the point marked by the dot, you're going to give it *one* sharp blow with a hammer. After hitting it, you want the puck to move from north to south at a speed similar to its initial west-to-east speed. Draw a force vector with its tail on the dot to show the direction in which you will aim your hammer blow.

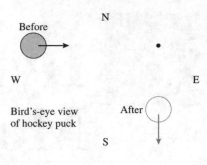

4. Tarzan swings through the jungle by hanging from a vine.
    a. Draw a motion diagram of Tarzan, as you learned in Chapter 1. Use it to find the direction of Tarzan's acceleration vector $\vec{a}$:
       i. Immediately after stepping off the branch, and
       ii. At the lowest point in his swing.

    b. At the lowest point in the swing, is the tension T in the vine greater than, less than, or equal to Tarzan's weight? Explain, basing your explanation on Newton's laws.

8.2 Uniform Circular Motion

5. The figure shows a *top view* of a plastic tube that is fixed on a horizontal tabletop. A marble is shot into the tube at A. On the figure, sketch the marble's trajectory after it leaves the tube at B.

Top view of
horizontal tube

6. A ball swings in a *vertical* circle on a string. During one revolution, a very sharp knife is used to cut the string at the instant when the ball is at its lowest point. Sketch the subsequent trajectory of the ball until it hits the ground.

7. The figures are a bird's-eye view of particles on a string moving in horizontal circles on a tabletop. All are moving at the same speed. Rank in order, from largest to smallest, the string tensions T_1 to T_4.

Order:

Explanation:

8. A ball on a string moves in a vertical circle. When the ball is at its lowest point, is the tension in the string greater than, less than, or equal to the ball's weight? Explain. (You may want to include a free-body diagram as part of your explanation.)

9. A marble rolls around the inside of a cone. Draw a free-body diagram of the marble when it is on the left side of the cone and a free-body diagram of the marble when it is on the right side of the cone.

On left side   On right side

10. Can the following be reasonably modeled as a central force with constant r? Answer Yes or No.

    a. A ball on a string is swung in a horizontal circle. _____

    b. A ball on a string is swung in a vertical circle. _____

    c. A satellite orbits the earth. _____

    d. A satellite spirals into a black hole. _____

    e. A car makes a U turn at constant speed. _____

    f. A child rides on a merry-go-round. _____

    g. An ant rides on a slowing turntable. _____

11. A coin of mass m is placed distance r from the center of a turntable. The coefficient of static friction
PSS between the coin and the turntable is μ_s. Starting from rest, the turntable is gradually rotated faster and
8.1 faster. At what angular velocity does the coin slip and "fly off"?

a. Begin with a pictorial representation. Draw the turntable both as seen from above and as an edge view with the coin on the left side coming toward you. Label radius r, make a table of known information, and indicate what you're trying to find.

b. What direction does $\vec{f}_s$ point? _____
Explain.

c. What condition describes the situation just as the coin starts to slip? Write this condition as a mathematical statement.

d. Now draw a free-body diagram of the coin. Following Problem Solving Strategy 8.1, draw the free-body diagram with the circle viewed edge on, the r-axis pointing toward the center of the circle, and the z-axis perpendicular to the plane of the circle. Your free-body diagram should have three forces on it.

e. Referring to Problem Solving Strategy 8.1, write Newton's second law for the r- and z-components of the forces. One sum should equal 0, the other mv^2/r.

f. The two equations of part e are valid for any angular velocity up to the point of slipping. If you combine these with your statement of part c, you can solve for the speed v_{max} at which the coin slips. Do so.

g. Finally, use the relationship between v and ω to find the angular velocity of slipping.

8.3 Circular Orbits

12. A small projectile is launched parallel to the ground at height $h = 1$ m with sufficient speed to orbit a completely smooth, airless planet. A bug rides in a small hole inside the projectile. Is the bug weightless? Explain.

8.4 Reasoning about Circular Motion

13. A stunt plane does a series of vertical loop the loops at a fairly steady speed. At what point in the circle does the pilot feel the heaviest? Explain. Include a free-body diagram with your explanation.

14. You can swing a ball on a string in a *vertical* circle if you swing it fast enough.

    a. Draw two free-body diagrams of the ball at the top of the circle. On the left, show the ball when it is going around the circle very fast. On the right, show the ball as it goes around the circle more slowly.

Very fast	Slower

    b. As you continue slowing the swing, there comes a frequency at which the string goes slack and the ball doesn't make it to the top of the circle. What condition must be satisfied for the ball to be able to complete the full circle?

    c. Suppose the ball has the smallest possible frequency that allows it to go all the way around the circle. What is the tension in the string when the ball is at the highest point? Explain.

15. It's been proposed that future space stations create "artificial gravity" by rotating around an axis.

   a. How would this work? Explain.

   b. Would the artificial gravity be equally effective throughout the space station? If not, where in the space station would the residents want to live and work?

8.5 Nonuniform Circular Motion

16. For each, figure determine the signs $(+$ or $-)$ of ω and α.

Speeding up

Slowing down

Slowing down

Speeding up

ω _____ ω _____ ω _____ ω _____

α _____ α _____ α _____ α _____

17. The figures below show the radial acceleration vector $\vec{a}_r$ at four sequential points on the trajectory of a particle moving in a counterclockwise circle.

   a. For each, draw the tangential acceleration vector $\vec{a}_t$ a at points 2 and 3 or, if appropriate, write a $\vec{a}_t = \vec{0}$.

   b. Determine whether a_t is positive $(+)$, negative $(-)$, or zero (0).

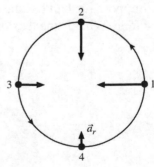

$a_t =$ _____ $a_t =$ _____ $a_t =$ _____

9 Work and Kinetic Energy

9.1 Energy Overview

1. Can the following be reasonably modeled with the basic energy model? Answer Yes or No.

   a. A box slides up and down a very smooth ramp. _____

   b. A baseball player slides into second base. _____

   c. A burner heats water to the boiling point. _____

   d. A burner heats a gas. The molecules move faster. _____

9.2 Work and Kinetic Energy for a Single Particle

2. On the axes below, draw graphs of the kinetic energy of
   a. A 1000 kg car that uniformly accelerates from 0 to 20 m/s in 20 s.
   b. A 1000 kg car moving at 20 m/s that brakes to a halt with uniform deceleration in 20 s.
   c. A 1000 kg car that drives once around a 130-m-diameter circle at a speed of 20 m/s.
   Calculate K at several times, plot the points, and draw a smooth curve between them.

a.

b.

c.

Exercises 3–10: For each situation:
- Draw a before-and-after pictorial representation.
- Draw and label the displacement vector $\Delta \vec{r}$.
- Draw a free-body diagram.
- Fill in the table by showing the sign ($+$, $-$, or 0) of the quantities listed.

3. An elevator moves up at constant speed.

W_{ten} _____

W_{grav} _____

W_{tot} _____

ΔK _____

4. A descending elevator brakes to a halt.

W_{tens} _____

W_{grav} _____

W_{tot} _____

ΔK _____

5. A box slides up a frictionless slope.

W_{norm} _____

W_{grav} _____

W_{tot} _____

ΔK _____

6. A rope pulls a box to the left across a frictionless floor.

W_{tens} _____

W_{norm} _____

W_{grav} _____

W_{tot} _____

ΔK _____

7. A ball is thrown straight up. Consider the ball from one microsecond after it leaves your hand until the highest point of its trajectory.

W_{hand} _____

W_{grav} _____

W_{tot} _____

ΔK _____

8. A car turns a corner at constant speed.

W_{fric} _____
W_{norm} _____
W_{grav} _____
W_{tot} _____
ΔK _____

9. A flat block on a string swings once around a horizontal circle on a frictionless table. The block moves at steady speed.

W_{tens} _____
W_{norm} _____
W_{grav} _____
W_{tot} _____
ΔK _____

10. A 0.2 kg plastic cart and a 20 kg lead cart both roll without friction on a horizontal surface. Equal forces are used to push both carts forward a distance of 1 m, starting from rest. After traveling 1 m, is the kinetic energy of the plastic cart greater than, less than, or equal to the kinetic energy of the lead cart? Explain.

11. Equal forces push together two equal-mass boxes on a frictionless surface. Both boxes have the same initial speed, and later both have the same slower speed. Let the system be the boxes and the spring.

a. Is the work done on the system positive, negative, or zero? Explain.

b. Is ΔE_{sys} positive, negative, or zero? Explain.

9.3 Calculating the Work Done

12. For each pair of vectors, is the sign of $\vec{A} \cdot \vec{B}$ positive (+), negative (−), or zero (0)?

a.

Sign = _____

b.

Sign = _____

c.

Sign = _____

d.

Sign = _____

e.

Sign = _____

f.

Sign = _____

13. Each of the diagrams below shows a vector $\vec{A}$. Draw and label a vector $\vec{B}$ that will cause $\vec{A} \cdot \vec{B}$ to have the sign indicated.

a.

$\vec{A} \cdot \vec{B} > 0$

b.

$\vec{A} \cdot \vec{B} < 0$

c.

$\vec{A} \cdot \vec{B} = 0$

14. If $\vec{A} \cdot \vec{B} = 0$, can you conclude that one of the vectors has zero magnitude. Explain?

15. Rudy picks up a 5 kg box and lifts it straight up, at constant speed, a height of 1 m. Beth uses a rope to pull a 5 kg box up a 15° frictionless slope, at constant speed, until it has reached a height of 1 m. Which of the two does more work? Or do they do equal amounts of work? Explain.

16. A sprinter running the 100-meter dash accelerates down the track. Is the work by the track on the sprinter positive, negative, or zero? Explain.

17. A particle moving along the x-axis experiences the forces shown below. How much work does each force do on the particle? What is each particle's change in kinetic energy?

a.

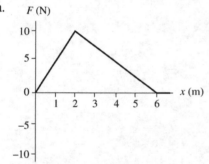

$W = $ _____

$\Delta K = $ _____

b.

$W = $ _____

$\Delta K = $ _____

18. A 1 kg particle moving along the x-axis experiences the force shown in the graph. If the particle's speed is 2 m/s at $x = 0$ m, what is its speed when it gets to $x = 5$ m?

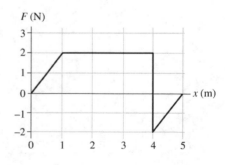

9.4 Restoring Forces and the Work Done by a Spring

19. A spring is attached to the floor and pulled straight up by a string. The string's tension is measured. The graph shows the tension in the string as a function of the spring's length L.

a. Does this spring obey Hooke's Law? Explain why or why not.

b. If it does, what is the spring constant?

20. Draw a figure analogous to Figure 9.17 in the textbook for a spring that is attached to a wall on the *right* end. Use the figure to show that F and Δs always have opposite signs.

21. A spring has an unstretched length of 10 cm. It exerts a restoring force F when stretched to a length of 11 cm.

    a. For what length of the spring is its restoring force $3F$?

    b. At what compressed length is the restoring force $2F$?

22. The left end of a spring is attached to a wall. When Bob pulls on the right end with a 200 N force, he stretches the spring by 20 cm. The same spring is then used for a tug-of-war between Bob and Carlos. Each pulls on his end of the spring with a 200 N force.

    a. How far does Bob's end of the spring move? Explain.

    b. How far does Carlos's end of the spring move? Explain.

23. In Example 9.9 in the textbook, a compressed spring with a spring constant of 65 N/m expands from $x_0 = -12$ cm $= -0.12$ m to its equilibrium position at $x_1 = 0$ m.

a. Graph the spring force $(F_{Sp})_s$ from $x_1 = -0.12$ m to $x_1 = 0$ m.

b. Use your graph to determine ΔK, the change in a cube's kinetic energy when launched by a spring that has been compressed by 12 cm.

c. Use your result from part b to find the launch speed of a 100 g cube in the absence of friction. Compare your answer to the value found in the Example 9.9.

9.5 Dissipative Forces and Thermal Energy

24. A ball of clay traveling at 10 m/s slams into a wall and sticks. What happens to the kinetic energy the clay had just before impact?

25. a. A baseball player slides into second base. What happens to the runner's kinetic energy?

    b. How should you define the system to analyze this situation with the energy principle?

9.6 Power

26. a. If you push an object 10 m with a 10 N force in the direction of motion, how much work do you do on it?

    b. How much power must you provide to push the object in 1 s? In 10 s? In 0.1 s?

27. You push with a force of 400 N and use 200 W of power to push a box across a rough floor at steady speed.
    a. How much force is needed to push the box twice as fast? Explain.

    b. How much power would you need to push the box twice as fast?

10 Interactions and Potential Energy

10.1 Potential Energy

10.2 Gravitational Potential Energy

1. Below we see a 1 kg object that is initially 1 m above the ground and rises to a height of 2 m. Anjay, Brittany, and Carlos each measure its position, but each of them uses a different coordinate system. Fill in the table to show the initial and final gravitational potential energies and ΔU as measured by our three aspiring scientists.

	U_i	U_f	ΔU
Anjay			
Brittany			
Carlos			

2. A roller coaster car rolls down a frictionless track, reaching speed v_f at the bottom.

   a. If you want the car to go twice as fast at the bottom, by what factor must you increase the height of the track?

   b. Does your answer to part a depend on whether the track is straight or not? Explain.

3. Below are shown three frictionless tracks. A ball is released from rest at the position shown on the left. To which point does the ball make it on the right before reversing direction and rolling back? Point B is the same height as the starting position.

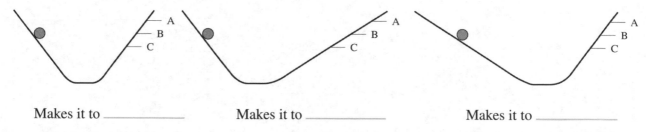

Makes it to _____   Makes it to _____   Makes it to _____

Exercises 4–6: Draw an energy bar chart to show the energy transformations for the situation described.

4. A car runs out of gas and coasts up a hill until finally stopping.

$$K_i \quad + \quad U_{Gi} \quad = \quad K_f \quad + \quad U_{Gf}$$

5. A pendulum is held out at 45° and released from rest. A short time later it swings through the lowest point on its arc.

$$K_i \quad + \quad U_{Gi} \quad = \quad K_f \quad + \quad U_{Gf}$$

6. A ball starts from rest on the top of one hill, rolls without friction through a valley, and just barely makes it to the top of an adjacent hill.

$$K_i \quad + \quad U_{Gi} \quad = \quad K_f \quad + \quad U_{Gf}$$

7. What energy transformations occur as a box slides up a gentle but slightly rough incline until stopping at the top?

10.3 Elastic Potential Energy

8. A heavy object is released from rest at position 1 above a spring. It falls and contacts the spring at position 2. The spring achieves maximum compression at position 3. Fill in the table below to indicate whether each of the quantities are $+$, $-$, or 0 during the intervals $1 \rightarrow 2$, $2 \rightarrow 3$, and $1 \rightarrow 3$.

	$1 \rightarrow 2$	$2 \rightarrow 3$	$1 \rightarrow 3$
ΔK			
ΔU_G			
ΔU_{Sp}			

9. Rank in order, from most to least, the amount of elastic potential energy $U_{Sp\,1}$ to $U_{Sp\,4}$ stored in each of these springs.

Order:

Explanation:

10. A spring gun shoots out a plastic ball at speed v_0 The spring is then compressed twice the distance it was on the first shot.

   a. By what factor is the spring's potential energy increased?

   b. By what factor is the ball's launch speed increased? Explain.

Exercises 11–12: Draw an energy bar chart to show the energy transformations for the situation described.

11. A bobsled sliding across frictionless, horizontal ice runs into a giant spring. A short time later the spring reaches its maximum compression.

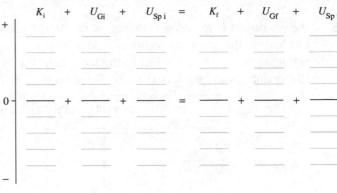

12. A brick is held above a spring that is standing on the ground. The brick is released from rest, and a short time later the spring reaches its maximum compression.

10.4 Conservation of Energy

13. Give a *specific* example of a situation in which:

    a. $W_{ext} \rightarrow K$ with $\Delta U = 0$ and $\Delta E_{th} = 0$.

    b. $W_{ext} \rightarrow U$ with $\Delta K = 0$ and $\Delta E_{th} = 0$.

    c. $W_{ext} \rightarrow E_{th}$ with $\Delta K = 0$ and $\Delta U = 0$.

14. A small cube of mass m slides back and forth in a frictionless, hemispherical
PSS bowl of radius R. Suppose the cube is released at angle θ. What is the cube's
10.1 speed at the bottom of the bowl?

a. Begin by drawing a before-and-after pictorial representation. Let the
cube's initial position and speed be y_i and v_i. Use a similar notation for the
final position and speed.

b. At the initial position, are either K_i or U_{Gi} zero? If so, which? _____

c. At the final position, are either K_f or U_{Gf} zero? If so, which? _____

d. Does thermal energy need to be considered in this situation? Why or why not?

e. Write the conservation of energy equation in terms of position and speed variables, omitting any
terms that are zero.

f. You're given not the initial position but the initial angle. Do the geometry and trigonometry to find y_i
in terms of R and θ.

g. Use your result of part f in the energy conservation equation, and then finish solving the problem.

10.5 Energy Diagrams

15. A particle with the potential energy shown in the graph is moving to the right at $x = 0$ m with total energy E.

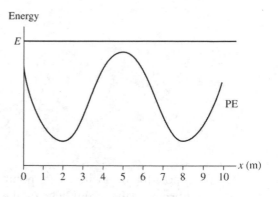

a. At what value or values of x is the particle's speed a maximum?

b. At what value or values of x is the particle's speed a minimum?

c. At what value or values of x is the potential energy a maximum?

d. Does this particle have a turning point in the range of x covered by the graph? If so, where?

16. The figure shows a potential-energy curve. Suppose a particle with total energy E_1 is at position A and moving to the right.

a. For each of the following regions of the x-axis, does the particle speed up, slow down, maintain a steady speed, or change direction?

A to B _____

B to C _____

C to D _____

D to E _____

E to F _____

b. Where is the particle's turning point? _____

c. For a particle that has total energy E_2 what are the possible motions and where do they occur along the x-axis?

d. What position or positions are points of stable equilibrium? For each, would a particle in equilibrium at that point have total energy $\leq E_2$, between E_2 and E_1, or $\geq E_1$?

e. What position or positions are points of unstable equilibrium? For each, would a particle in equilibrium at that point have total energy $\leq E_2$, between E_2 and E_1, or $\geq E_1$?

17. Below are a set of axes on which you are going to draw a potential-energy curve. By doing experiments, you find the following information:

   • A particle with energy E_1 oscillates between positions D and E.
   • A particle with energy E_2 oscillates between positions C and F.
   • A particle with energy E_3 oscillates between positions B and G.
   • A particle with energy E_4 enters from the right, bounces at A, then never returns.

   Draw a potential-energy curve that is consistent with this information.

10.6 Force and Potential Energy

10.7 Conservative and Nonconservative Forces

18. The graph shows the potential-energy curve of a particle moving along the *x*-axis under the influence of a conservative force.

    a. In which intervals of *x* is the force on the particle to the right?

    b. In which intervals of *x* is the force on the particle to the left?

    c. At what value or values of *x* is the magnitude of the force a maximum?

    d. What value or values of *x* are positions of stable equilibrium?

    e. What value or values of *x* are positions of unstable equilibrium?

    f. If the particle is released from rest at *x* = 0 m, will it reach *x* = 10 m? Explain.

19. a. If the force on a particle at some point in space is zero, must its potential energy also be zero at that point? Explain.

    b. If the potential energy of a particle at some point in space is zero, must the force on it also be zero at that point? Explain.

10.8 The Energy Principle Revisited

20. A system loses 1000 J of potential energy. In the process, it does 500 J of work on the environment and the thermal energy increases by 250 J. Show this process on an energy bar chart.

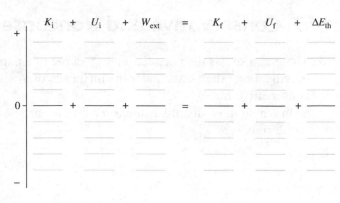

21. A system gains 1000 J of kinetic energy while losing 500 J of potential energy. The thermal energy increases by 250 J. Show this process on an energy bar chart.

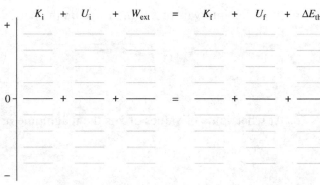

22. A box is sitting at the top of a ramp. An external force pushes the box down the ramp, causing it to slowly accelerate. Show this process on an energy bar chart.

23. Can the following be reasonably modeled with the basic energy model? Answer Yes or No.

   a. A car skids to a halt. _____

   b. A car accelerates away from a stop sign. _____

   c. A person lifts a weight. _____

   d. A compressed spring launches a weight. _____

   e. A burner heats a gas that expands and lifts a weight. _____

11 Impulse and Momentum

11.1 Momentum and Impulse

1. Rank in order, from largest to smallest, the momenta $(p_x)_1$ to $(p_x)_5$.

Order:

2. The position-versus-time graph is shown for a 500 g object. Draw the corresponding momentum-versus-time graph. Supply an appropriate scale on the vertical axis.

3. The momentum-versus-time graph is shown for a 500 g object. Draw the corresponding acceleration-versus-time graph. Supply an appropriate scale on the vertical axis.

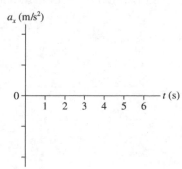

4. A 2 kg object is moving to the right with a speed of 1 m/s when it experiences an impulse due to the force shown in the graph. What is the object's speed and direction after the impulse?

a.

b.

5. A 2 kg object is moving to the left with a speed of 1 m/s when it experiences an impulse due to the force shown in the graph. What is the object's speed and direction after the impulse?

a.

b.

6. A 2 kg object has the velocity graph shown.

   a. What is the object's initial momentum? _____

   b. What is the object's final momentum? _____

   c. What impulse does the object experience? _____

   d. Draw the graph showing the force on the object.

7. A carnival game requires you to knock over a wood post by throwing a ball at it. You're offered a very bouncy rubber ball and a very sticky clay ball of equal mass. Assume that you can throw them with equal speed and equal accuracy. You get only one throw.

   a. Which ball will you choose? Why?

   b. Let's think about the situation more carefully. Both balls have the same initial momentum p_{ix} just before hitting the post. The clay ball sticks, the rubber ball bounces off with essentially no loss of speed. In terms of p_{ix}, what is the final momentum of each ball?

   Clay ball: $p_{fx} =$ _____   Rubber ball: $p_{fx} =$ _____

   Hint: Momentum has a sign. Did you take the sign into account?

   c. What is the *change* in the momentum of each ball?

   Clay ball: $\Delta p_x =$ _____   Rubber ball: $\Delta p_x =$ _____

   d. Which ball experiences a larger impulse during the collision? Explain.

   e. From Newton's third law, the impulse that the ball exerts on the post is equal in magnitude, although opposite in direction, to the impulse that the post exerts on the ball. Which ball exerts the larger impulse on the post?

   f. Don't change your answer to part a, but are you still happy with that answer? If not, how would you change your answer? Why?

8. Automobiles are designed with "crumple zones" intended to collapse in a collision. Use the ideas and language of this chapter to explain why.

9. A small, light ball S and a large, heavy ball L move toward each other, collide, and bounce apart.

a. Compare the force that S exerts on L during the collision to the force that L exerts on S. That is, is $F_{\text{S on L}}$ larger, smaller, or equal to $F_{\text{L on S}}$? Explain.

b. Compare the time interval during which S experiences a force to the time interval during which L experiences a force. Are they equal, or is one longer than the other?

c. Sketch a graph showing a *plausible* $F_{\text{S on L}}$ as a function of time and another graph showing a plausible $F_{\text{L on S}}$ as a function of time. Be sure to think about the *sign* of each force.

d. Compare the impulse delivered to S to the impulse delivered to L. Explain.

e. Compare the momentum change of S to the momentum change of L.

f. Compare the velocity change of S to the velocity change of L.

g. What is the change in the *sum* of the momenta of the two balls? Is it positive, negative, or zero?

Exercises 10–12: Draw a momentum bar chart to show the momenta and impulse for the situation described.

10. A compressed spring shoots a ball to the right. The ball was initially at rest.

$$+ \quad | \quad p_{ix} \quad + \quad J_x \quad = \quad p_{fx}$$

11. A rubber ball is tossed straight up and bounces off the ceiling. Consider only the collision with the ceiling.

$$+ \quad | \quad p_{iy} \quad + \quad J_y \quad = \quad p_{fy}$$

12. A clay ball is tossed straight up and sticks to the ceiling. Consider only the collision with the ceiling.

$$+ \quad | \quad p_{iy} \quad + \quad J_y \quad = \quad p_{fy}$$

13. Particle A has less mass than particle B. Both are pushed forward across a frictionless surface by equal forces for 1 s. Both start from rest.

   a. Compare the amount of work done on each particle. That is, is the work done on A greater than, less than, or equal to the work done on B? Explain.

   b. Compare the impulses delivered to particles A and B. Explain.

   c. Compare the final speeds of particles A and B. Explain.

11.2 Conservation of Momentum

14. A golf club continues forward after hitting the golf ball. Is momentum conserved in the collision? Explain, making sure you are careful to identify the "system."

15. As you release a ball, it falls—gaining speed and momentum. Is momentum conserved?
    a. Answer this question from the perspective of choosing the ball alone as the system.

    b. Answer this question from the perspective of choosing ball + earth as the system.

16. Two particles collide, one of which was initially moving and the other initially at rest.
    a. Is it possible for *both* particles to be at rest after the collision? Give an example in which this happens, or explain why it can't happen.

    b. Is it possible for *one* particle to be at rest after the collision? Give an example in which this happens, or explain why it can't happen.

17. A tennis ball traveling to the left at speed v_{Bi} is hit by a tennis racket moving to the right at speed v_{Ri}.

Although the racket is swung in a circular arc, its forward motion during the collision with the ball is so small that we can consider it to be moving in a straight line. Further, we can invoke the *impulse approximation* to neglect the steady force of the arm on the racket during the brief duration of its collision with the ball. Afterward, the ball is returned to the right at speed v_{Bf}. What is the racket's speed after it hits the ball? The masses of the ball and racket are m_B and m_R, respectively.

   a. Begin by drawing a before-and-after pictorial representation as described in Tactics Box 9.1. You can assume that the racket continues in the forward direction but at a reduced speed.

   b. Define the system. That is, what object or objects should be inside the system so that it is an *isolated system* whose momentum is conserved?

   c. Write an expression for P_{ix}, the total momentum of the system before the collision. Your expression should be written using the quantities given in the problem statement. Notice, however, that you're given *speeds*, but momentum is defined in terms of *velocities*. Based on your coordinate system and the directions of motion, you may need to give a negative momentum to one or more objects.

   d. Now write an expression for P_{fx}, the total momentum of the system after the collision.

   e. If you chose the system correctly, its momentum is conserved. So equate your expressions for the initial and final total momentum, and then solve for what you want to find.

11.3 Collisions

11.4 Explosions

Exercises 18–20: Prepare a pictorial representation for these problems, but *do not* solve them.
- Draw pictures of "before" and "after."
- Define symbols relevant to the problem.
- List known information, and identify the desired unknown.

18. A 50 kg archer, standing on frictionless ice, shoots a 100 g arrow at a speed of 100 m/s. What is the recoil speed of the archer?

19. The parking brake on a 2000 kg Cadillac has failed, and it is rolling slowly, at 1 mph, toward a group of small innocent children. As you see the situation, you realize there is just time for you to drive your 1000 kg Volkswagen head-on into the Cadillac and thus to save the children. With what speed should you impact the Cadillac to bring it to a halt?

20. Dan is gliding on his skateboard at 4 m/s. He suddenly jumps backward off the skateboard, kicking the skateboard forward at 8 m/s. How fast is Dan going as his feet hit the ground? Dan's mass is 50 kg and the skateboard's mass is 5 kg.

21. Ball 1 with an initial speed of 14 m/s has a perfectly elastic collision with ball 2 that is initially at rest. Afterward, the speed of ball 2 is 21 m/s.

   a. What will be the speed of ball 2 if the initial speed of ball 1 is doubled?

   b. What will be the speed of ball 2 if the mass of ball 1 is doubled?

22. Indicate whether each of the following can be reasonably modeled as a perfectly elastic collision, a perfectly inelastic collision, both, or neither.

   a. A bowling ball hits a bowling pin. _____

   b. A bug hits your windshield while you're driving. _____

   c. A bat hits a Styrofoam ball. _____

   d. A flying bird meets and swallows a flying bug. _____

   e. A dropped ball bounces back to half its initial height. _____

11.5 Momentum in Two Dimensions

23. An object initially at rest explodes into three fragments. The momentum vectors of two of the fragments are shown. Draw the momentum vector $\vec{p}_3$ of the third fragment.

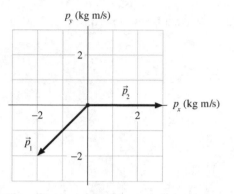

24. An object initially at rest explodes into three fragments. The momentum vectors of two of the fragments are shown. Draw the momentum vector $\vec{p}_3$ of the third fragment.

25. A 500 g ball traveling to the right at 8.0 m/s collides with and bounces off another ball. The figure shows the momentum vector $\vec{p}_1$ of the first ball after the collision. Draw the momentum vector $\vec{p}_2$ of the second ball.

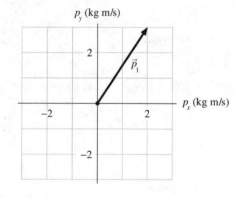

26. A 500 g ball traveling to the right at 4.0 m/s collides with and bounces off another ball. The figure shows the momentum vector $\vec{p}_1$ of the first ball after the collision. Draw the momentum vector $\vec{p}_2$ of the second ball.

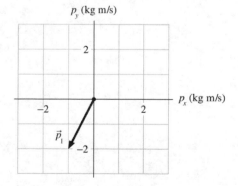

11.6 Rocket Propulsion

27. Write a brief description of how a rocket accelerates through the vacuum of space.

28. a. If a rocket motor's exhaust speed is increased by a factor of 2, does the rocket's maximum speed increase by more than a factor or 2, a factor of 2, or less than a factor of 2? Explain.

    b. If a rocket's fuel-to-empty-rocket mass ratio is increased by a factor of 2, does the rocket's maximum speed increase by more than a factor or 2, a factor of 2, or less than a factor of 2? Explain.

12 Rotation of a Rigid Body

12.1 Rotational Motion

1. Can the following be reasonably modeled as rigid bodies? Answer Yes or No.

    a. A car wheel and tire. _____

    b. A person. _____

    c. A yo-yo. _____

    d. A bowl of Jello. _____

2. The following figures show a wheel rolling on a ramp. Determine the signs ($+$ or $-$) of the wheel's angular velocity and angular acceleration.

    ω _____ ω _____ ω _____
    α _____ α _____ α _____

3. A ball is rolling back and forth inside a bowl. The figure shows the ball at extreme left edge of the ball's motion as it changes direction.

    a. At this point, is ω positive, negative, or zero? _____

    b. At this point, is α positive, negative, or zero? _____

4. Point B on a rotating wheel is twice as far from the axle as point A.

    a. Is ω_B equal to $\frac{1}{2}\omega_A$, ω_A, or $2\omega_A$? Explain.

    b. Is v_B equal to $\frac{1}{2}v_A$, v_A, or $2v_A$? Explain.

5. A wheel rolls to the left along a horizontal surface, up a ramp, then continues along the upper horizontal surface. Draw graphs for the wheel's angular velocity ω and angular acceleration α a as a function of time.

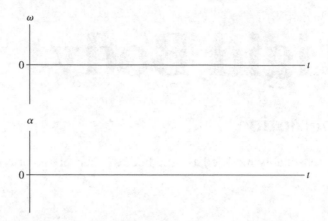

6. A wheel rolls to the right along the surface shown. Draw graphs for the wheel's angular velocity ω and angular acceleration α until the wheel reaches its highest point on the right side.

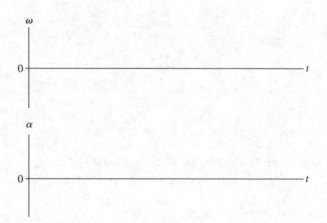

12.2 Rotation about the Center of Mass

7. Is the center of mass of this dumbbell at point 1, 2, or 3? Explain.

8. Mark the center of mass of this object with an ×.

12.3 Rotational Energy

12.4 Calculating Moment of Inertia

9. The figure shows four equal-mass bars rotating about their center. Rank in order, from largest to smallest, their rotational kinetic energies K_1 to K_4.

Order:

Explanation:

10. Two solid spheres have the same mass. Sphere B has twice the rotational kinetic energy of sphere A.

   a. What is the ratio R_B/R_A of their radii?

   b. Would your answer change if both spheres were hollow? Explain.

   c. Would your answer change if A were solid and B were hollow? Explain.

11. Which has more kinetic energy: a particle of mass M rotating with angular velocity ω in a circle of radius R, or a sphere of mass M and radius R spinning at angular velocity ω? Explain.

12. The moment of inertia of a uniform rod about an axis through its center is $\frac{1}{12}ML^2$. The moment of inertia about an axis at one end is $\frac{1}{3}ML^2$. Explain *why* the moment of inertia is larger about the end than about the center.

13. You have two steel spheres. Sphere 2 has three times the radius of sphere 1. By what *factor* does the moment of inertia I_2 of sphere 2 exceed the moment of inertia I_1 of sphere 1?

14. The professor hands you two spheres. They have the same mass, the same radius, and the same exterior surface. The professor claims that one is a solid sphere and that the other is hollow. Can you determine which is which without cutting them open? If so, how? If not, why not?

15. Rank in order, from largest to smallest, the moments of inertia I_1, I_2, and I_3 about the midpoint of the rod.

Order:

Explanation:

12.5 Torque

16. Five forces are applied to a door. For each, determine if the torque about the hinge is positive $(+)$, negative $(-)$, or zero (0).

$\tau_1=$_____ $\tau_3=$_____

$\tau_5=$_____

$\tau_2=$_____ $\tau_4=$_____

17. Six forces, each of magnitude either F or $2F$, are applied to a door. Rank in order, from largest to smallest, the six torques τ_1 to τ_6 about the hinge.

   Order:

   Explanation:

18. A bicycle is at rest on a smooth surface. A force is applied to the bottom pedal as shown. Does the bicycle roll forward (to the right), backward (to the left), or not at all? Explain.

19. Four forces are applied to a rod that can pivot on an axle. For each force,

    a. Use a **black** pen or pencil to draw the line of action.
    b. Use a **red** pen or pencil to draw and label the moment arm, or state that $d = 0$.
    c. Determine if the torque about the axle is positive $(+)$, negative $(-)$ or zero (0). Write your answer in the blank.

20. a. Draw a force vector at A whose torque about the axle is negative.
    b. Draw a force vector at B whose torque about the axle is zero.
    c. Draw a force vector at C whose torque about the axle is positive.

21. The dumbbells below are all the same size, and the forces all have the same magnitude. Rank in order, from largest to smallest, the torques τ_1, τ_2, and τ_3 about the midpoint of the rod.

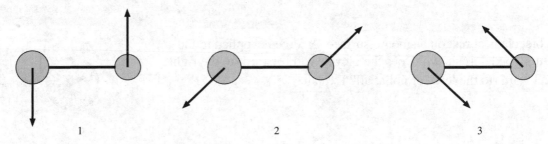

Order:

Explanation:

12.6 Rotational Dynamics

22. A student gives a quick push to a ball at the end of a massless, rigid rod, causing the ball to rotate clockwise in a *horizontal* circle. The rod's pivot is frictionless.

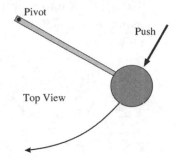

a. As the student is pushing, is the torque about the pivot positive, negative, or zero?

b. After the push has ended, does the ball's angular velocity
   i. Steadily increase?
   ii. Increase for awhile, then hold steady?
   iii.Hold steady?
   iv. Decrease for awhile, then hold steady?
   v. Steadily decrease?
   Explain the reason for your choice.

c. Right after the push has ended, is the torque positive, negative, or zero?

23. a. Rank in order, from largest to smallest, the torques τ_1 to τ_4 about the center of the wheel.

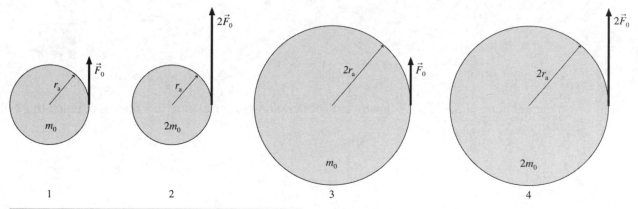

Order:
Explanation:

b. Rank in order, from largest to smallest, the angular accelerations α_1 to α_4.

24. The top graph shows the torque on a rotating wheel as a function of time. The wheel's moment of inertia is 10 kg m^2. Draw graphs of α-versus-t and ω-versus-t, assuming $\omega_0 = 0$. Provide units and appropriate scales on the vertical axes.

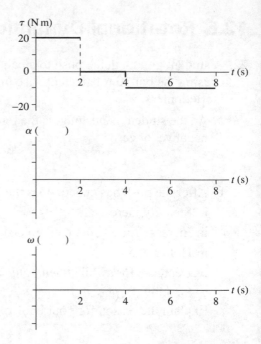

25. The wheel turns on a frictionless axle. A string wrapped around the smaller diameter shaft is tied to a block. The block is released at $t = 0$ s and hits the ground at $t = t_1$.

a. Draw a graph of ω-versus-t for the wheel, starting at $t = 0$ s and continuing to some time $t > t_1$.

b. Is the magnitude of the block's downward acceleration greater than g, less than g, or equal to g? Explain.

12.7 Rotation about a Fixed Axis

26. A square plate can rotate about an axle through its center. Four forces of equal magnitude are applied, one at a time, to different points on the plate. The forces turn as the plate rotates, maintaining the same orientation with respect to the plate. Rank in order, from largest to smallest, the angular accelerations α_1 to α_4 caused by the four forces.

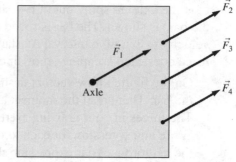

Order:

Explanation:

27. A solid cylinder and a cylindrical shell have the same mass, same radius, and turn on frictionless, horizontal axles. (The cylindrical shell has light-weight spokes connecting the shell to the axle.) A rope is wrapped around each cylinder and tied to a block. The blocks have the same mass and are held the same height above the ground. Both blocks are released simultaneously. The ropes do not slip.

Which block hits the ground first? Or is it a tie? Explain.

28. A metal bar of mass M and length L can rotate in a horizontal
PSS plane about a vertical, frictionless axle through its center.
12.1 A hollow channel down the bar allows compressed air (fed
    in at the axle) to spray out of two small holes at the ends of
    the bar, as shown. The bar is found to speed up to angular
    velocity ω in a time interval Δt, starting from rest. What
    force does each escaping jet of air exert on the bar?

Top view

   a. On the figure, draw vectors to show the forces exerted on
      the bar. Then label the moment arms of each force.

   b. The forces in your drawing exert a torque about the axles.
      Write an expression for each torque, and then add them to get
      the net torque. Your expression should be in terms of the unknown force F and "known" quantities
      such as M, L, g, etc.

   c. What is the moment of inertia of this bar about the axle? _____

   d. According to Newton's second law, the torque causes the bar to undergo an angular acceleration.
      Use your results from parts b and c to write an expression for the angular acceleration. Simplify the
      expression as much as possible.

   e. You can now use rotational kinematics to write an expression for the bar's angular velocity after time
      Δt has elapsed. Do so.

   f. Finally, solve your equation in part e for the unknown force.

      This is now a result you could use with experimental measurements to determine the size of the
      force exerted by the gas.

12.8 Static Equilibrium

29. A uniform rod pivots about a frictionless, horizontal axle through its center. It is placed on a stand, held motionless in the position shown, then gently released. On the right side of the figure, draw the final, equilibrium position of the rod. Explain your reasoning.

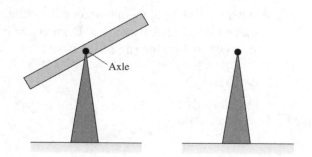

30. The dumbbell has masses m and $2m$. Force $\vec{F}_1$ acts on mass m in the direction shown. Is there a force $\vec{F}_2$ that can act on mass $2m$ such that the dumbbell moves with pure translational motion, without any rotation? If so, draw $\vec{F}_2$, making sure that its length shows the magnitude of $\vec{F}_2$ relative to $\vec{F}_1$. If not, explain why not.

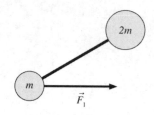

31. Forces $\vec{F}_1$ and $\vec{F}_2$ have the same magnitude and are applied to the corners of a square plate. Is there a *single* force $\vec{F}_3$ that, if applied to the appropriate point on the plate, will cause the plate to be in total equilibrium? If so, draw it, making sure it has the right position, orientation, and length. If not, explain why not.

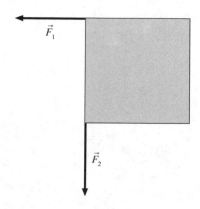

12.9 Rolling Motion

32. A wheel is rolling along a horizontal surface with the center-of-mass velocity shown. Draw the velocity vector $\vec{v}$ at points 1 to 4 on the rim of the wheel.

12.10 The Vector Description of Rotational Motion

12.11 Angular Momentum

33. For each vector pair $\vec{A}$ and $\vec{B}$ shown below, determine if $\vec{A} \times \vec{B}$ points into the page, out of the page, or is zero.

$\vec{A} \times \vec{B}$ _____ $\vec{A} \times \vec{B}$ _____ $\vec{A} \times \vec{B}$ _____ $\vec{A} \times \vec{B}$ _____

34. Each figure below shows $\vec{A}$ and $\vec{A} \times \vec{B}$. Determine if $\vec{B}$ is in the plane of the page or perpendicular to the page. If $\vec{B}$ is in the plane of the page, draw it. If $\vec{B}$ is perpendicular to the page, state whether $\vec{B}$ points into the page or out of the page.

$\vec{B}$ _____ $\vec{B}$ _____ $\vec{B}$ _____

35. Draw the angular velocity vector on each of the rotating wheels.

a.   b.   c.

36. The figures below show a force acting on a particle. For each, draw the torque vector for the torque about the origin.
   • Place the tail of the torque vector at the origin.
   • Draw the vector large and straight (use a ruler!) so that its direction is clear. Use dotted lines from the tip of the vector to the axes to show the plane in which the vector lies.

a.   b.   c.

37. The figures below show a particle with velocity $\vec{v}$. For each, draw the angular momentum vector $\vec{L}$ for the angular momentum relative to the origin. Place the tail of the angular momentum vector at the origin.

a.

b.

c.

38. Rank in order, from largest to smallest, the angular momenta L_1 to L_4.

1   2   3   4

Order:

Explanation:

39. Disks 1 and 2 have equal mass. Is the angular momentum of disk 2 larger than, smaller than, or equal to the angular momentum of disk 1? Explain.

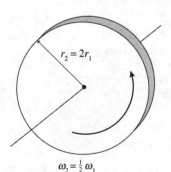

13 Newton's Theory of Gravity

13.1 A Little History

13.2 Isaac Newton

13.3 Newton's Law of Gravity

1. Is the earth's gravitational force on the moon larger than, smaller than, or equal to the moon's gravitational force on the earth? Explain.

2. Star A is twice as massive as star B. They attract each other.

   a. Draw gravitational force vectors on both stars. The length of each vector should be proportional to the size of the force.

   b. Is the acceleration of star A larger than, smaller than, or equal to the acceleration of star B? Explain.

3. The gravitational force of a star on orbiting planet 1 is F_1. Planet 2, which is twice as massive as planet 1 and orbits at half the distance from the star, experiences gravitational force F_2. What is the ratio F_2/F_1?

4. Comets orbit the sun in highly elliptical orbits. A new comet is sighted at time t_1.

   Comet's orbit
   Sun

   a. Later, at time t_2, the comet's acceleration a_2 is twice as large as the acceleration a_1 it had at t_1. What is the ratio r_2/r_1 of the comet's distance from the sun at t_2 to its distance at t_1?

   b. Still later, at time t_3, the comet has rounded the sun and is headed back out to the farthest reaches of the solar system. The size of the force F_3 on the comet at t_3 is the same as the size of force F_2 at t_2, but the comet's distance from the sun r_3 is only 90% of distance r_2. Astronomers recognize that the comet has lost mass. Part of it was "boiled away" by the heat of the sun during the time of closest approach, thus forming the comet's tail. What percent of its initial mass did the comet lose?

13.4 Little *g* and Big *G*

5. How far away from the earth does an orbiting spacecraft have to be in order for the astronauts inside to be weightless?

6. The free-fall acceleration at the surface of planet 1 is 20 m/s^2. The radius and the mass of planet 2 are half those of planet 1. What is g on planet 2?

13.5 Gravitational Potential Energy

7. Explain *why* the gravitational potential energy of two masses is negative. Note that saying "because that's what the formula gives" is *not* an explanation. An explanation makes use of the basic ideas of force and potential energy.

13.6 Satellite Orbits and Energies

8. Planet X orbits the star Alpha with a "year" that is 200 earth days long. Planet Y circles Alpha at nine times the distance of planet X. How long is a year on planet Y?

9. The mass of Jupiter is $M_{Jupiter} = 300 M_{earth}$. Jupiter orbits around the sun with $T_{Jupiter} = 11.9$ years in an orbit with $r_{Jupiter} = 5.2 r_{earth}$. Suppose the earth could be moved to the distance of Jupiter and placed in a circular orbit around the sun. The new period of the earth's orbit would be

   a. 1 year.   b. 11.9 years.

   c. Between 1 year and 11.9 years.   d. More than 11.9 years.

   e. It could be anything, depending on   f. It is impossible for a planet of earth's mass
      the speed the earth is given.   to orbit at the distance of Jupiter.

   Circle the letter of the true statement. Then explain your choice.

10. Satellite A orbits a planet with a speed of 10,000 m/s. Satellite B is twice as massive as satellite A and orbits at twice the distance from the center of the planet. What is the speed of satellite B?

11. a. A crew of a spacecraft in a clockwise circular orbit around the moon wants to change to a new orbit that will take them down to the surface. In which direction should they fire the rocket engine? On the figure, show the exhaust gases coming out of the spacecraft.

    b. On the figure, show the spacecraft's orbit after firing its rocket engine.

    c. The moon has no atmosphere, so the spacecraft will continue unimpeded along its new orbit until either firing its rocket again or (ouch!) intersecting the surface. As it descends, does its speed increase, decrease, or stay the same? Explain.

14 Fluids and Elasticity

14.1 Fluids

1. An object has density ρ.

   a. Suppose each of the object's three dimensions is increased by a factor of 2 without changing the material of which the object is made. Will the density change? If so, by what factor? Explain.

   b. Suppose each of the object's three dimensions is increased by a factor of 2 without changing the object's mass. Will the density change? If so, by what factor? Explain.

2. Air enclosed in a cylinder has density $\rho = 1.4 \text{ kg/m}^3$.

   a. What will be the density of the air if the length of the cylinder is doubled while the radius is unchanged?

   b. What will be the density of the air if the radius of the cylinder is halved while the length is unchanged?

3. Air enclosed in a sphere has density $\rho = 1.4 \text{ kg/m}^3$. What will the density be if the radius of the sphere is halved?

14.2 Pressure

14.3 Measuring and Using Pressure

4. When you stand on a bathroom scale, it reads 700 N. Suppose a giant vacuum cleaner sucks half the air out of the room, reducing the pressure to 0.5 atm. Would the scale reading increase, decrease, or stay the same? Explain.

5. Rank in order, from largest to smallest, the pressures at A, B, and C.

   Order:

   Explanation:

6. Refer to the figure in Exercise 5. Rank in order, from largest to smallest, the pressures at D, E, and F.

   Order:

   Explanation:

7. The gauge pressure at the bottom of a cylinder of liquid is $p_g = 0.4$ atm. The liquid is poured into another cylinder with twice the radius of the first cylinder. What is the gauge pressure at the bottom of the second cylinder?

8. Cylinders A and B contain liquids. The pressure p_A at the bottom of A is higher than the pressure p_B at the bottom of B. Is the ratio p_A/p_B of the absolute pressures larger than, smaller than, or equal to the ratio of the gauge pressures? Explain.

9. A and B are rectangular tanks full of water. They have equal depths, equal thicknesses (the dimension into the page), but different widths.

   a. Compare the forces the water exerts on the bottoms of the tanks. Is F_A larger than, smaller than, or equal to F_B? Explain.

   A   Sides   B

   b. Compare the forces the water exerts on the sides of the tanks. Is F_A larger than, smaller than, or equal to F_B? Explain.

10. Water expands when heated. Suppose a beaker of water is heated from 10°C to 90°C. Does the pressure at the bottom of the beaker increase, decrease, or stay the same? Explain.

11. Is p_A larger than, smaller than, or equal to p_B? Explain.

    A   B

12. It is well known that you can trap liquid in a drinking straw by placing the tip of your finger over the top while the straw is in the liquid, and then lifting it out. The liquid runs out when you release your finger.

    a. What is the *net* force on the cylinder of trapped liquid?

    b. Three forces act on the trapped liquid. Draw and label all three on the figure.

    c. Is the gas pressure inside the straw, between the liquid and your finger, greater than, less than, or equal to atmospheric pressure? Explain, basing your explanation on your answers to parts a and b.

    d. If your answer to part c was "greater" or "less," how did the pressure change from the atmospheric pressure that was present when you placed your finger over the top of the straw?

13. At sea level, the height of the mercury column in a sealed glass tube is 380 mm. What can you say about the contents of the space above the mercury? Be as specific as you can.

    380 mm

14.4 Buoyancy

14. Rank in order, from largest to smallest, the densities of A, B, and C.

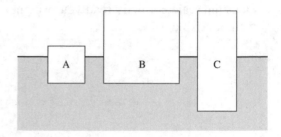

Order:
Explanation:

15. A, B, and C have the same volume. Rank in order, from largest to smallest, the sizes of the buoyant forces F_A, F_B, and F_C on A, B, and C.

Order:
Explanation:

16. Refer to the figure of Exercise 15. Now A, B, and C have the same density. Rank in order, from largest to smallest, the sizes of the buoyant forces on A, B, and C.

Order:
Explanation:

17. Suppose you stand on a bathroom scale that is on the bottom of a swimming pool. The water comes up to your waist.

    Does the scale correctly read your weight *mg*? If not, does the scale read more than or less than your weight? Explain.

18. Ships A and B have the same height and the same mass. Their cross-section profiles are shown in the figure. Does one ship ride higher in the water (more height above the water line) than the other? If so, which one? Explain.

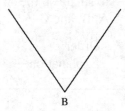

A   B

14.5 Fluid Dynamics

19. A stream flows from left to right through the constant-depth channel shown below in an overhead view. A grid has been added to facilitate measurement. The fluid's flow speed at A is 2 m/s.

a. Shade in squares to represent the water that has flowed past point A in the last two seconds.

b. Shade in squares to represent the water that has flowed past point B in the last two seconds.

20. Can the following be reasonably modeled as ideal fluids? Answer Yes or No.

a. Water flowing through a tube. _____

b. Honey flowing through a tube. _____

c. A whitewater river. _____

d. Water flowing slowing through a concrete-lined channel. _____

21. Liquid flows through a pipe. You can't see into the pipe to know how the inner diameter changes. Rank in order, from largest to smallest, the flow speeds v_1 to v_3 at points 1, 2, and 3.

Order:
Explanation:

22. Wind blows over a house. A window on the ground floor is open. Is there an air flow through the house? If so, does the air flow in the window and out the chimney, or in the chimney and out the window? Explain.

Wind

14.6 Elasticity

23. A force stretches a wire by 1 mm.

    a. A second wire of the same material has the same cross section and twice the length. How far will it be stretched by the same force? Explain.

    b. A third wire of the same material has the same length and twice the diameter as the first. How far will it be stretched by the same force? Explain.

24. A 2000 N force stretches a wire by 1 mm.

    a. A second wire of the same material is twice as long and has twice the diameter. How much force is needed to stretch it by 1 mm? Explain.

    b. A third wire is twice as long as the first and has the same diameter. How far is it stretched by a 4000 N force?

25. A wire is stretched right to the breaking point by a 5000 N force. A longer wire made of the same material has the same diameter. Is the force that will stretch it right to the breaking point larger than, smaller than, or equal to 5000 N? Explain.

26. Sphere A is compressed by 1% at an ocean depth of 4000 m. Sphere B is compressed by 1% at an ocean depth of 5000 m. Which has the larger bulk modulus? Explain.

15 Oscillations

15.1 Simple Harmonic Motion

1. Give three examples of *oscillatory* motion. (Note that circular motion is not the same as oscillatory motion.)

2. On the axes below, sketch three cycles of the displacement-versus-time graph for:

   a. A particle undergoing symmetric periodic motion that is *not* SHM.

   b. A particle undergoing asymmetric periodic motion.

   c. A particle undergoing simple harmonic motion.

3. Consider the particle whose motion is represented by the *x*-versus-*t* graph below.

a. Is this periodic motion? _____   b. Is this motion SHM? _____

c. What is the period? _____   d. What is the frequency? _____

e. You learned in Chapter 2 to relate velocity graphs to position graphs. Use that knowledge to draw the particle's velocity-versus-time graph on the axes provided.

4. Shown below is the velocity-versus-time graph of a particle.

a. What is the period of the motion? _____

b. Draw the particle's position-versus-time graph for an oscillation around $x = 0$.

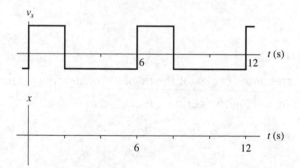

5. The figure shows the position-versus-time graph of a particle in SHM.

a. At what times is the particle moving to the right at maximum speed?

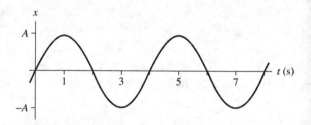

b. At what times is the particle moving to the left at maximum speed?

c. At what times is the particle instantaneously at rest?

15.2 SHM and Circular Motion

6. A particle goes around a circle 5 times at constant speed, taking a total of 2.5 seconds.

   a. Through what angle *in degrees* has the particle moved? _____

   b. Through what angle *in radians* has the particle moved? _____

   c. What is the particle's frequency f?

   d. Use your answer to part b to determine the particle's angular frequency ω.

   e. Does ω (in rad/s) $= 2\pi f$(in Hz)? _____

7. A particle moves counterclockwise around a circle at constant speed. For each of the phase constants given below:

   • Show with a dot *on the circle* the particle's starting position.
   • Sketch two cycles of the particle's *x*-versus-*t* graph.

a.

b.

c.

d.

8. a. On the top set of axes below, sketch two cycles of the *x*-versus-*t* graphs for a particle in simple harmonic motion with phase constants i) $\phi_0 = \pi/2$ rad and ii) $\phi_0 = -\pi/2$ rad.

   b. Use the bottom set of axes to sketch velocity-versus-time graphs for the particles. Make sure each velocity graph aligns vertically with the correct points on the *x*-versus-*t* graph.

9. The graph below represents a particle in simple harmonic motion.

   a. What is the phase constant ϕ_0? Explain how you determined it.

   b. What is the phase of the particle at each of the three numbered points on the graph?

   Phase at 1: _____   Phase at 2: _____   Phase at 3: _____

   c. Place dots on the circle above to show the position of a circular-motion particle at the times corresponding to points 1, 2, and 3. Label each dot with the appropriate number.

15.3 Energy in SHM

10. The figure shows the potential-energy diagram of a particle oscillating on a spring.

    a. What is the spring's equilibrium length?

    _____

    b. The particle's turning points are at 14 cm and 26 cm. Draw the total energy line and label it TE.

    c. What is the particle's maximum kinetic energy?

    _____

    d. Draw a graph of the particle's kinetic energy as a function of position.

    e. What will be the turning points if the particle's total energy is doubled?

    _____

11. A block oscillating on a spring has an amplitude of 20 cm. What will be the block's amplitude if its total energy is tripled? Explain.

    _____

12. A block oscillating on a spring has a maximum speed of 20 cm/s. What will be the block's maximum speed if its total energy is tripled? Explain.

    _____

13. The figure shows the potential energy diagram of a particle.

    a. Is the particle's motion periodic? How can you tell?

    b. Is the particle's motion simple harmonic motion? How can you tell?

    c. What is the amplitude of the motion?

    d. Draw a graph of the particle's kinetic energy as a function of position.

14. Equation 15.25 in the textbook states that $\frac{1}{2}kA^2 = \frac{1}{2}mv_{max}^2$. What does this mean? Write a couple of sentences explaining how to interpret this equation.

15.4 The Dynamics of SHM

15.5 Vertical Oscillations

15. A block oscillating on a spring has period $T = 4$ s.
    a. What is the period if the block's mass is halved? Explain.
       Note: You do not know values for either m or k. Do *not* assume any particular values for them. The required analysis involves thinking about ratios.

    b. What is the period if the value of the spring constant is quadrupled?

    c. What is the period if the oscillation amplitude is doubled while m and k are unchanged?

16. For graphs a and b, determine:
    - The angular frequency ω.
    - The oscillation amplitude A.
    - The phase constant ϕ_0.

    Note: Graphs a and b are independent. Graph b is *not* the velocity graph of a.

a.

b.

$\omega =$ _____ $\omega =$ _____
$A =$ _____ $A =$ _____
$\phi_0 =$ _____ $\phi_0 =$ _____

17. The graph on the right is the position-versus-time graph for a simple harmonic oscillator.

   a. Draw the v_x-versus-t and a_x-versus-t graphs.

   b. When x is greater than zero, is a_x ever greater than zero? If so, at which points in the cycle?

   c. When x is less than zero, is a_x ever less than zero? If so, at which points in the cycle?

   d. Can you make a general conclusion about the relationship between the sign of x and the sign of a_x?

   e. When x is greater than zero, is v_x ever greater than zero? If so, how is the oscillator moving at those times?

18. For the oscillation shown on the left below:

   a. What is the phase constant ϕ_0? _____

   b. Draw the corresponding v_x-versus-t graph on the axes below the x-versus-t graph.

   c. On the axes on the right, sketch two cycles of the x-versus-t and the v_x-versus-t graphs if the value of ϕ_0 found in part a is replaced by its negative, $-\phi_0$.

   d. Describe *physically* what is the same and what is different about the initial conditions for two oscillators having "equal but opposite" phase constants ϕ_0 and $-\phi_0$.

19. The top graph shows the position versus time for a mass oscillating on a spring. On the axes below, sketch the position-versus-time graph for this block for the following situations:

    Note: The changes described in each part refer back to the original oscillation, not to the oscillation of the previous part of the question. Assume that all other parameters remain constant. Use the same horizontal and vertical scales as the original oscillation graph.

    a. The amplitude and the frequency are doubled.

    b. The amplitude is halved and the mass is quadrupled.

    c. The phase constant is increased by $\pi/2$ rad.

    d. The maximum speed is doubled while the amplitude remains constant.

15.6 The Pendulum

20. A pendulum on planet X, where the value of g is unknown, oscillates with a period of 2 seconds. What is the period of this pendulum if:

   a. Its mass is tripled?

   Note: You do not know the values of m, L, or g, so do not assume any specific values.

   b. Its length is tripled?

   c. Its oscillation amplitude is tripled?

21. The graph shows the displacement s versus time for an oscillating pendulum.

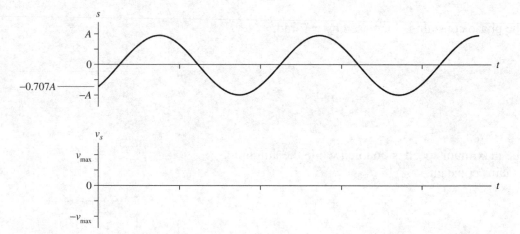

   a. Draw the pendulum's velocity-versus-time graph.
   b. What is the value of the phase constant ϕ_0?

   c. In the space at the right, draw a *picture* of the pendulum that shows (and labels!)

   • The extremes of its motion.
   • Its position at $t = 0$ s.
   • Its direction of motion (using an arrow) at $t = 0$ s.

15.7 Damped Oscillations

22. Can the following be reasonably modeled as SHM? Answer Yes or No.

    a. A perfectly elastic ball bouncing up and down to the same height. _____

    b. A marble rolling in the bottom of a bowl. _____

    c. A plastic ruler clamped at one end and "plucked" at the other. _____

    d. A heavy mass hanging from a rope that is twisting back and forth. _____

23. If the damping constant b of an oscillator is increased,

    a. Is the medium more resistive or less resistive? _____

    b. Do the oscillations damp out more quickly or less quickly? _____

    c. Is the time constant τ increased or decreased? _____

24. A block on a spring oscillates horizontally on a table with friction. Draw and label force vectors on the block to show all *horizontal* forces on the block.

    a. The mass is to the right of the equilibrium point and moving away from it.

    b. The mass is to the right of the equilibrium point and approaching it.

25. A mass oscillating on a spring has a frequency of 0.5 Hz and a damping time constant $\tau = 5$ s. Use the grid below to draw a reasonably accurate position-versus-time graph lasting 40 s.

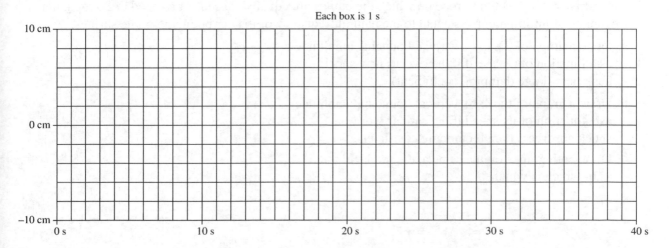

26. The figure below shows the envelope of the oscillations of a lightly damped oscillator. On the same axes, draw the envelope of oscillations if
    a. The time constant is doubled.
    b. The time constant is halved.

15.8 Driven Oscillations and Resonance

27. What is the difference between the driving frequency and the natural frequency of an oscillator?

28. A car drives along a bumpy road on which the bumps are equally spaced. At a speed of 20 mph, the frequency of hitting bumps is equal to the natural frequency of the car bouncing on its springs.

    a. Draw a graph of the car's vertical bouncing amplitude as a function of its speed if the car has new shock absorbers (large damping coefficient).

    b. Draw a graph of the car's vertical bouncing amplitude as a function of its speed if the car has worn-out shock absorbers (small damping coefficient).

    Draw both graphs on the same axes, and label them as to which is which.

16 Traveling Waves

16.1 An Introduction to Waves

1. a. In your own words, define what a *transverse wave* is.

    b. Give an example of a wave that, from your own experience, you know is a transverse wave. What observations or evidence tells you this is a transverse wave?

2. a. In your own words, define what a *longitudinal wave* is.

    b. Give an example of a wave that, from your own experience, you know is a longitudinal wave. What observations or evidence tells you this is a longitudinal wave?

3. Three wave pulses travel along the same string. Rank in order, from largest to smallest, their wave speeds v_1, v_2, and v_3.

    Order:

    Explanation:

16.2 One-Dimensional Waves

4. A wave pulse travels along a string at a speed of 200 cm/s. What will be the speed if:
   Note: Each part below is independent and refers to changes made to the original string.

   a. The string's tension is doubled?

   b. The string's mass is quadrupled (but its length is unchanged)?

   c. The string's length is quadrupled (but its mass is unchanged)?

   d. The string's mass and length are both quadrupled?

5. This is a history graph showing the displacement as a function of time at one point on a string. Did the displacement at this point reach its maximum of 2 mm *before* or *after* the interval of time when the displacement was a constant 1 mm? Explain how you interpreted the graph to answer this question.

6. Each figure below shows a snapshot graph at time $t = 0$ s of a wave pulse on a string. The pulse on the left is traveling to the right at 100 cm/s; the pulse on the right is traveling to the left at 100 cm/s. Draw snapshot graphs of the wave pulse at the times shown next to the axes.

a.

b.

7. This snapshot graph is taken from Exercise 6a. On the axes below, draw the *history* graphs $D(x = 2 \text{ cm}, t)$ and $D(x = 6 \text{ cm}, t)$, showing the displacement at $x = 2$ cm and $x = 6$ cm as functions of time. Refer to your graphs in Exercise 6a to see what is happening at different instants of time.

8. This snapshot graph is from Exercise 6b.

   a. Draw the history graph $D(x = 0$ cm, $t)$ for this wave at the point $x = 0$ cm.

   b. Draw the *velocity*-versus-time graph for the piece of the string at $x = 0$ cm. Imagine painting a dot on the string at $x = 0$ cm. What is the velocity of this dot as a function of time as the wave passes by?

   c. As a wave passes through a medium, is the speed of a particle in the medium the same as or different from the speed of the wave through the medium? Explain.

9. Below are four snapshot graphs of wave pulses on a string. For each, draw the history graph at the specified point on the x-axis. No time scale is provided on the t-axis, so you must determine an appropriate time scale and label the t-axis appropriately.

   a.

   At $t = 0$ s

   100 cm/s

   x (cm)

   At $x = 2$ cm

   t

   b.

   At $t = 0$ s

   100 cm/s

   x (cm)

   At $x = 8$ cm

   t

   c.

   At $t = 0.02$ s

   100 cm/s

   x (cm)

   At $x = 0$ cm

   t

   d.

   At $t = 0.02$ s

   x (cm)

   100 cm/s

   At $x = 0$ cm

   t

10. A history graph $D(x = 0 \text{ cm}, t)$ is shown for the $x = 0$ cm point on a string. The pulse is moving to the right at 100 cm/s.

a. Does the $x = 0$ cm point on the string rise quickly and then fall slowly, or rise slowly and then fall quickly? Explain.

b. At what time does the leading edge of the wave pulse arrive at $x = 0$ cm? _____

c. At $t = 0$ s, how far is the leading edge of the wave pulse from $x = 0$ cm? Explain.

d. At $t = 0$ s, is the leading edge to the right or to the left of $x = 0$ cm? _____

e. At what time does the trailing edge of the wave pulse leave $x = 0$ cm? _____

f. At $t = 0$ s, how far is the trailing edge of the pulse from $x = 0$ cm? Explain.

g. By referring to the answers you've just given, draw a snapshot graph $D(x, t = 0 \text{ s})$ showing the wave pulse on the string at $t = 0$ s.

11. These are a history graph *and* a snapshot graph for a wave pulse on a string. They describe the same wave from two perspectives.

a. In which direction is the wave traveling? Explain.

b. What is the speed of this wave?

12. Below are two history graphs for wave pulses on a string. The speed and direction of each pulse are indicated. For each, draw the snapshot graph at the specified instant of time. No distance scale is provided, so you must determine an appropriate scale and label the x-axis appropriately.

a.

100 cm/s to the left

b.

50 cm/s to the left

Snapshot at $t = -0.02$ s

Snapshot at $t = 0.06$ s

13. A horizontal Slinky is at rest on a table. A wave pulse is sent along the Slinky, causing the top of link 5 to move *horizontally* with the displacement shown in the graph.

   a. Is this a transverse or a longitudinal wave? Explain.

   b. What is the position of link 5 at $t = 0.1$ s? _____

      What is the position of link 5 at $t = 0.2$ s? _____

      What is the position of link 5 at $t = 0.3$ s? _____

      Note: *Position*, not displacement.

   c. Draw a velocity-versus-time graph of link 5. Add an appropriate scale to the vertical axis.

   d. Can you determine, from the information given, whether the wave pulse is traveling to the right or to the left? If so, give the direction and explain how you found it. If not, why not?

   e. Can you determine, from the information given, the speed of the wave? If so, give the speed and explain how you found it. If not, why not?

14. We can use a series of dots to represent the positions of the links in a Slinky. The top set of dots shows a Slinky in equilibrium with a 1 cm spacing between the links. A wave pulse is sent down the Slinky, traveling to the right at 10 cm/s. The second set of dots shows the Slinky at $t = 0$ s. The links are numbered, and you can measure the displacement Δx of each link.

a. Draw a snapshot graph showing the displacement of each link at $t = 0$ s. There are 13 links, so your graph should have 13 dots. Connect your dots with lines to make a continuous graph.

b. Is your graph a "picture" of the wave or a "representation" of the wave? Explain.

c. Draw graphs of displacement versus the link number at $t = 0.1$ s and $t = 0.2$ s.

d. Now draw dot pictures of the links at $t = 0.1$ s and $t = 0.2$ s. The equilibrium positions and the $t = 0$ s picture are shown for reference.

15. The graph shows displacement versus the link number for a wave pulse on a Slinky. Draw a dot picture showing the Slinky at this instant of time. A picture of the Slinky in equilibrium, with 1 cm spacings, is given for reference.

16. The graph shows displacement versus the link number for a wave pulse on a Slinky. Draw a dot picture showing the Slinky at this instant of time. A picture of the Slinky in equilibrium, with 1 cm spacings, is given for reference.

16.3 Sinusoidal Waves

17. The figure shows a sinusoidal traveling wave. Draw a graph of the wave if:

a. Its amplitude is halved and its wavelength is doubled.

b. Its speed is doubled and its frequency is quadrupled.

18. The wave shown at time $t = 0$ s is traveling to the right at a speed of 25 cm/s.

a. Draw snapshot graphs of this wave at times $t = 0.1$ s, $t = 0.2$ s, $t = 0.3$ s, and $t = 0.4$ s.

b. What is the wavelength of the wave?

c. Based on your graphs, what is the period of the wave?

d. What is the frequency of the wave?

e. What is the value of the product λf?

f. How does this value of λf compare to the speed of the wave?

19. Three waves traveling to the right are shown below. The first two are shown at $t = 0$, the third at $t = T/2$. What are the phase constants ϕ_0 of these three waves?

$\phi_0 =$ _____ $\phi_0 =$ _____ $\phi_0 =$ _____

Note: Knowing the displacement $D(0, 0)$ is a *necessary* piece of information for finding ϕ_0 but is not by itself enough. The first two waves above have the same value for $D(0, 0)$ but they do *not* have the same ϕ_0. You must also consider the overall shape of the wave.

20. A sinusoidal wave with wavelength 2 m is traveling along the x-axis. At $t = 0$ s the wave's phase at $x = 2$ m is $\pi/2$ rad.

    a. Draw a snapshot graph of the wave at $t = 0$ s.

    b. At $t = 0$ s, what is the phase at $x = 0$ m? _____
    c. At $t = 0$ s, what is the phase at $x = 1$ m? _____
    d. At $t = 0$ s, what is the phase at $x = 3$ m? _____

    Note: No calculations are needed. Think about what the phase *means* and utilize your graph.

21. Consider the wave shown. Redraw this wave if:

    a. Its wave number is doubled.

    b. Its wave number is halved.

16.4 The Wave Equation on a String

No exercises

16.5 Sound and Light

22. Rank in order, from largest to smallest, the wavelengths λ_1 to λ_3 for sound waves having frequencies $f_1 = 100$ Hz, $f_2 = 1000$ Hz, and $f_3 = 10{,}000$ Hz.

> Order:
>
> Explanation:

23. A light wave travels from vacuum, through a transparent material, and back to vacuum. What is the index of refraction of this material? Explain.

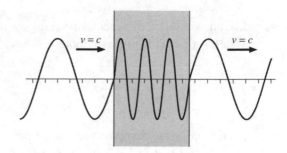

24. A light wave travels from vacuum, through a transparent material whose index of refraction is $n = 2.0$, and back to vacuum. Finish drawing the snapshot graph of the light wave at this instant.

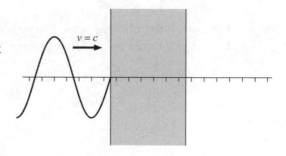

16.6 The Wave Equation in a Fluid

25. At 2°C, the speed of sound in water is 1410 m/s.

    a. What would be the speed of sound in a liquid with the same density as water but twice the bulk modulus?

    b. What would be the speed of sound in a liquid with the same bulk modulus as water but twice the density?

16.7 Waves in Two and Three Dimensions

26. A wave-front diagram is shown for a sinusoidal plane wave at time $t = 0$ s. The diagram shows only the xy-plane, but the wave extends above and below the plane of the paper.

a. What is the wavelength of this wave? _____

b. At $t = 0$ s, for which values of y is the wave a crest?

c. At $t = 0$ s, for which values of y is the wave a trough?

d. Can you tell if this is a transverse or a longitudinal wave? If so, which is it and how did you determine it? If not, why not?

e. How does the displacement at the point $(x, y, z) = (6, 5, 0)$ compare to the displacement at the point $(2, 5, 0)$? Is it more, less, the same, or is there no way to tell? Explain.

f. On the left axes below, draw a snapshot graph $D(y, t = 0 \text{ s})$ along the y-axis at $t = 0$ s.

g. On the right axes below, draw a wave-front diagram at time $t = 0.3$ s.

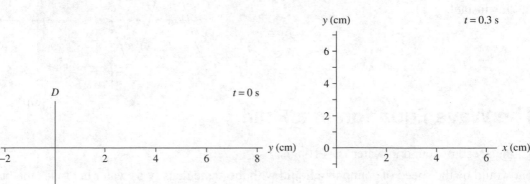

27. These are the wave fronts of a circular wave. What is the phase difference between:

a. Points A and B? _____

b. Points C and D? _____

c. Points E and F? _____

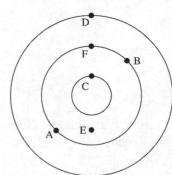

16.8 Power, Intensity, and Decibels

28. A laser beam has intensity I_0.

    a. What is the intensity, in terms of I_0, if a lens focuses the laser beam to $\frac{1}{10}$ its initial diameter?

    b. What is the intensity, in terms of I_0, if a lens defocuses the laser beam to 10 times its initial diameter?

29. Sound wave A delivers 2 J of energy in 2 s. Sound wave B delivers 10 J of energy in 5 s. Sound wave C delivers 2 mJ of energy in 1 ms. Rank in order, from largest to smallest, the sound powers P_A, P_B, and P_C of these three sound waves.

    Order:

    Explanation:

30. A giant chorus of 1000 male vocalists is singing the same note. Suddenly 999 vocalists stop, leaving one soloist. By how many decibels does the sound intensity level decrease? Explain.

16.9 The Doppler Effect

31. You are standing at $x = 0$ m, listening to a sound that is emitted at frequency f_0. The graph shows the frequency you hear during a four-second interval. Which of the following describes the sound source?

    i.   It moves from left to right and passes you at $t = 2$ s.
    ii. It moves from right to left and passes you at $t = 2$ s.
    iii. It moves toward you but doesn't reach you. It then reverses direction at $t = 2$ s.
    iv. It moves away from you until $t = 2$ s. It then reverses direction and moves toward you but doesn't reach you.

    Explain your choice.

32. You are standing at $x = 0$ m, listening to a sound that is emitted at frequency f_0. At $t = 0$ s, the sound source is at $x = 20$ m and moving toward you at a steady 10 m/s. Draw a graph showing the frequency you hear from $t = 0$ s to $t = 4$ s. Only the shape of the graph is important, not the numerical values of f.

33. You are standing at $x = 0$ m, listening to seven identical sound sources. At $t = 0$ s, all seven are at $x = 343$ m and moving as shown below. The sound from all seven will reach your ear at $t = 1$ s.

    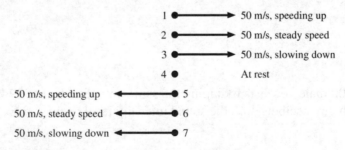

    Rank in order, from highest to lowest, the seven frequencies f_1 to f_7 that you hear at $t = 1$ s.

    Order:

    Explanation:

17 Superposition

17.1 The Principle of Superposition

1. Two pulses on a string, both traveling at 10 m/s, are approaching each other. Draw snapshot graphs of the string at the three times indicated.

2. Two pulses on a string, both traveling at 10 m/s, are approaching each other. Draw a snapshot graph of the string at $t = 1$ s.

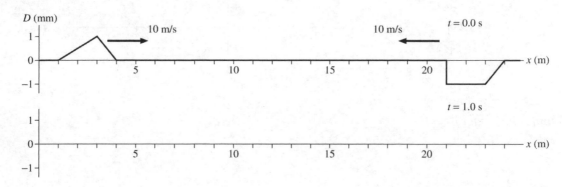

17.2 Standing Waves

17.3 Standing Waves on a String

3. Two waves are traveling in opposite directions along a string. Each has a speed of 1 cm/s and an amplitude of 1 cm. The first set of graphs below shows each wave at $t = 0$ s.

   a. On the axes at the right, draw the superposition of these two waves at $t = 0$ s.

   b. On the axes at the left, draw each of the two displacements every 2 s until $t = 8$ s. The waves extend beyond the graph edges, so new pieces of the wave will move in.

   c. On the axes at the right, draw the superposition of the two waves at the same instant.

(Continues next page)

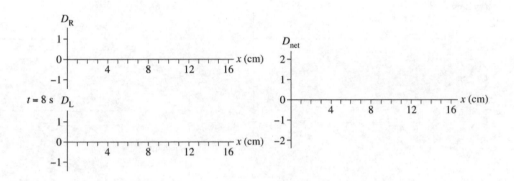

4. The figure shows a standing wave on a string.

a. Draw the standing wave if the tension is quadrupled while the frequency is held constant.

Original wave, tension T Tension $4T$

b. Suppose the tension is merely doubled while the frequency remains constant. Will there be a standing wave? If so, how many antinodes will it have? If not, why not?

5. This standing wave has a period of 8 ms. Draw snapshot graphs of the string every 1 ms from $t = 1$ ms to $t = 8$ ms. Think carefully about the proper amplitude at each instant.

6. The figure shows a standing wave on a string. It has frequency f.

   a. Draw the standing wave if the frequency is changed to $\frac{2}{3}f$ and to $\frac{3}{2}f$.

   Original wave, frequency f Frequency $\frac{2}{3}f$ Frequency $\frac{3}{2}f$

   b. Is there a standing wave if the frequency is changed to $\frac{1}{4}f$? If so, how many antinodes does it have? If not, why not?

17.4 Standing Sound Waves and Musical Acoustics

7. The picture shows a displacement standing sound wave in a 32-mm-long tube of air that is open at both ends.

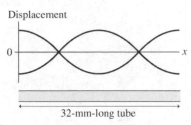

Displacement

32-mm-long tube

a. Which mode (value of m) standing wave is this? _____

b. Are the air molecules vibrating vertically or horizontally? Explain.

c. At what distances from the left end of the tube do the molecules oscillate with maximum amplitude?

8. The purpose of this exercise is to visualize the motion of the air molecules for the standing wave of Exercise 7. On the next page are nine graphs, every one-eighth of a period from $t = 0$ to $t = T$. Each graph represents the displacements at that instant of time of the molecules in a 32-mm-long tube. Positive values are displacements to the right; negative values are displacements to the left.

a. Consider nine air molecules that, in equilibrium, are 4 mm apart and lie along the axis of the tube. The top picture on the right shows these molecules in their equilibrium positions. The dotted lines down the page—spaced 4 mm apart—are reference lines showing the equilibrium positions. Read each graph carefully, then draw nine dots to show the positions of the nine air molecules at each instant of time. The first one, for $t = 0$, has already been done to illustrate the procedure.

   Note: It's a good approximation to assume that the left dot moves in the pattern 4, 3, 0, −3, −4, −3, 0, 3, 4 mm; the second dot in the pattern 3, 2, 0, −2, −3, −2, 0, 2, 3 mm; and so on.

b. At what times does the air reach maximum compression, and where does it occur?

   Max compression at time _____   Max compression at position _____

c. What is the relationship between the positions of maximum compression and the nodes of the standing wave?

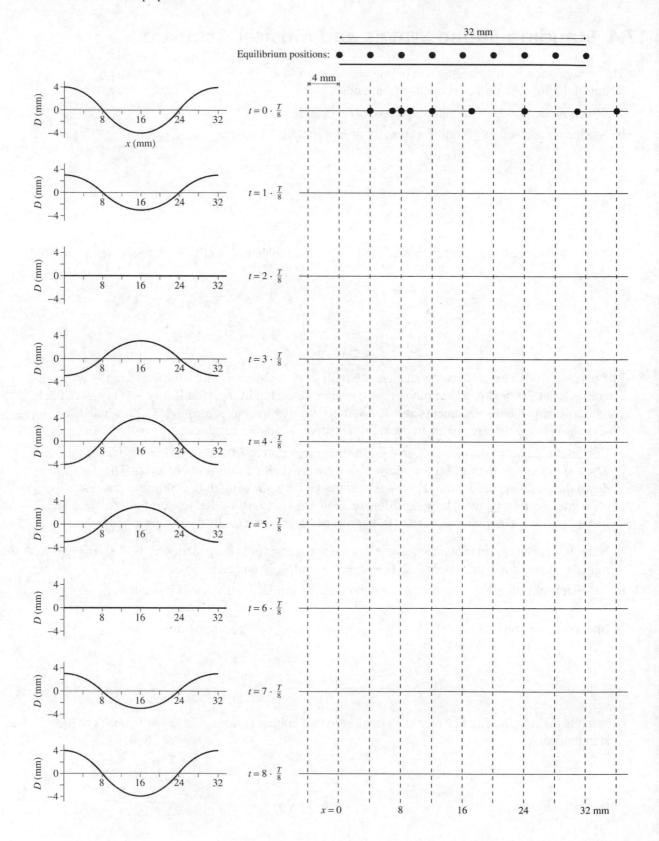

17.5 Interference in One Dimension

17.6 The Mathematics of Interference

9. The figure shows a snapshot graph at $t = 0$ s of loudspeakers emitting triangular-shaped sound waves. Speaker 2 can be moved forward or backward along the axis. Both speakers vibrate in phase at the same frequency. The second speaker is drawn below the first, so that the figure is clear, but you want to think of the two waves as overlapped as they travel along the x-axis.

   a. On the left set of axes, draw the $t = 0$ s snapshot graph of the second wave if speaker 2 is placed at each of the positions shown. The first graph, with $x_{\text{speaker}} = 2$ m, is already drawn.

   b. On the right set of axes, draw the superposition $D_{\text{net}} = D_1 + D_2$ of the waves from the two speakers. D_{net} exists only to the right of *both* speakers. It is the net wave traveling to the right.

   c. What separations between the speakers give constructive interference? _____

   d. What are the $\Delta x/\lambda$ ratios at the points of constructive interference? _____

   e. What separations between the speakers give destructive interference? _____

   f. What are the $\Delta x/\lambda$ ratios at the points of destructive interference? _____

10. Two loudspeakers are shown at $t = 0$ s. Speaker 2 is 4 m behind speaker 1.

$t = 0$ s

a. What is the wavelength λ? _____

b. Is the interference constructive or destructive?

_____

c. What is the phase constant ϕ_{10} for wave 1? _____

What is the phase constant ϕ_{20} for wave 2? _____

d. At points A, B, C, and D on the x-axis, what are:
   • The distances x_1 and x_2 to the two speakers?
   • The path length difference $\Delta x = x_2 - x_1$?
   • The phases ϕ_1 and ϕ_2 of the two waves at the point (not the phase constant)?
   • The phase difference $\Delta\phi = \phi_2 - \phi_1$?

Point A is already filled in to illustrate.

	x_1	x_2	Δx	ϕ_1	ϕ_2	$\Delta\phi$
Point A	1 m	5 m	4 m	0.5π rad	2.5π rad	2π rad
Point B						
Point C						
Point D						

e. Now speaker 2 is placed 2 m behind speaker 1.
   Is the interference constructive or destructive?

_____

$t = 0$ s

f. Repeat step d for the points A, B, C, and D.

_____

	x_1	x_2	Δx	ϕ_1	ϕ_2	$\Delta\phi$
Point A	1 m	3 m	2 m	0.5π rad	1.5π rad	π rad
Point B						
Point C						
Point D						

g. When the interference is constructive, what is $\Delta x/\lambda$? _____ What is $\Delta\phi$? _____

h. When the interference is destructive, what is $\Delta x/\lambda$? _____ What is $\Delta\phi$? _____

11. Two speakers are placed side-by-side at $x = 0$ m. The waves are shown at $t = 0$ s.

a. Is the interference constructive or destructive?

b. What is the phase constant ϕ_{10} for wave 1? _____

What is the phase constant ϕ_{10} for wave 2? _____

c. At points A, B, C, and D on the x-axis, what are:
   • The distances x_1 and x_2 to the two speakers?
   • The path length difference $\Delta x = x_2 - x_1$?
   • The phases ϕ_1 and ϕ_2 of the two waves at the point (not the phase constant)?
   • The phase difference $\Delta\phi = \phi_2 - \phi_1$?

	x_1	x_2	Δx	ϕ_1	ϕ_2	$\Delta\phi$
Point A						
Point B						
Point C						
Point D						

d. Speaker 2 is moved back 2 m. Does this change its phase constant ϕ_0?

e. Is the interference constructive or destructive?

f. Repeat step c for the points A, B, C, and D.

	x_1	x_2	Δx	ϕ_1	ϕ_2	$\Delta\phi$
Point A						
Point B						
Point C						
Point D						

12. Review your answers to the Exercises 10 and 11. Is it the separation path length difference Δx or the phase difference $\Delta\phi$ between the waves that determines whether the interference is constructive or destructive? Explain.

17.7 Interference in Two and Three Dimensions

13. This is a snapshot graph of two plane waves passing through a region of space. Each has a 2 mm amplitude. At each lettered point, what are the displacements of each wave and the net displacement?

a. Point A: $D_1 =$ _____ $D_2 =$ _____ $D_{net} =$ _____

b. Point B: $D_1 =$ _____ $D_2 =$ _____ $D_{net} =$ _____

c. Point C: $D_1 =$ _____ $D_2 =$ _____ $D_{net} =$ _____

14. Speakers 1 and 2 are 12 m apart. Both emit identical triangular sound waves with $\lambda = 4$ m and $\phi = \pi/2$ rad. Point A is $r_1 = 16$ m from speaker 1.

a. What is distance r_2 from speaker 2 to A? _____

b. Draw the wave from speaker 2 along the dashed line to just past point A.

c. At A, is wave 1 a crest, trough, or zero? _____

At A, is wave 2 a crest, trough, or zero? _____

d. What is the path length difference $\Delta r = r_2 - r_1$? _____ What is the ratio $\Delta r/\lambda$? _____

e. Is the interference at point A constructive, destructive, or in between? _____

15. Speakers 1 and 2 are 18 m apart. Both emit identical triangular sound waves with $\lambda = 4$ m and $\phi_0 = \pi/2$ rad. Point B is $r_1 = 24$ m from speaker 1.

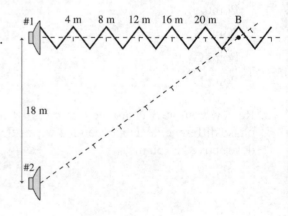

a. What is distance r_2 from speaker 2 to B? _____

b. Draw the wave from speaker 2 along the dashed line to just past point A.

c. At B, is wave 1 a crest, trough, or zero? _____

At B, is wave 2 a crest, trough, or zero? _____

d. What is the path length difference $\Delta r = r_2 - r_1$? _____ What is the ratio $\Delta r/\lambda$? _____

e. Is the interference at point B constructive, destructive, or in between? _____

16. Two speakers 12 m apart emit identical triangular sound waves with $\lambda = 4$ m and $\phi_0 = 0$ rad. The distances r_1 to points A, B, C, D, and E are shown in the table below.

a. For each point, fill in the table and determine whether the interference is constructive (C) or destructive (D).

Point	r_1	r_2	Δr	$\Delta r/\lambda$	C or D
A	2.2 m				
B	5.0 m				
C	9.0 m				
D	16 m				
E	35 m				

b. Are there any points to the right of E, on the line straight out from speaker 1, for which the interference is either exactly constructive or exactly destructive? If so, where? If not, why not?

c. Suppose you start at speaker 1 and walk straight away from the speaker for 50 m. Describe what you will hear as you walk.

17. The figure shows the wave-front pattern emitted by two loudspeakers.

 a. Draw a dot • at points where there is constructive interference. These will be points where two crests overlap *or* two troughs overlap.

 b. Draw an open circle o at points where there is destructive interference. These will be points where a crest overlaps a trough.

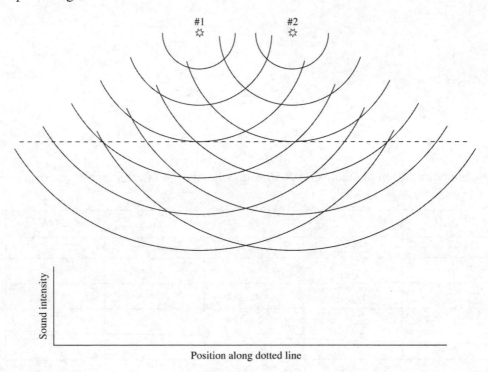

 c. Use a **black** line to draw each antinodal line of constructive interference. Use a **red** line to draw each nodal line of destructive interference.

 d. Draw a graph on the axes above of the sound intensity you would hear if you walked along the horizontal dashed line. Use the same horizontal scale as the figure so that your graph lines up with the figure above it.

 e. Suppose the phase constant of speaker 2 is increased by π rad. Describe what will happen to the interference pattern.

18. Two identical, in-phase loudspeakers are in a plane, distance d apart. Both emit sound waves with
PSS wavelength λ, with $\lambda \ll d$. At what distances r_1 directly in front of speaker 1 is there maximum
17.1 constructive interference of the two sound waves?

a. Begin with a visual representation. Draw two
speakers, one directly above the other, on the
left edge of the empty space. Label them 1
and 2, and show the distance between them
as d. Draw a horizontal line straight out from
speaker 1. Place a dot on this line, and label
its distance from the speaker as r_1. Under
what conditions is the interference at this
point constructive?

b. Referring to PSS 17.1, write the condition for constructive interference for two in-phase sources.
Then rearrange it to be of the form $\Delta r = \ldots$

c. Draw and label r_2 on your diagram. Write an expression for r_2 in terms of r_1 and d.

d. The path-length difference is $\Delta r = r_2 - r_1$. Use your part c result to rewrite the condition, from
part b, for maximum constructive interference.

e. Solve this equation for r_1. Do so by first isolating the square-root term on one side, then squaring
both sides.

f. Is there a solution for $m = 0$? _____ Does this make sense? $m = 0$ corresponds to zero path-
length difference. Is there any point on the line you drew where the path-length difference would be
zero? Explain.

g. Write explicit results for the values of r_1 giving constructive interference with $m = 1$ and $m = 2$. Which of these is closer to speaker 1? Explain why this is so.

h. What is the maximum value of m for which there is a meaningful solution? What's going on physically that prevents m from exceeding this value?

17.8 Beats

19. The two waves arrive simultaneously at a point in space from two different sources.

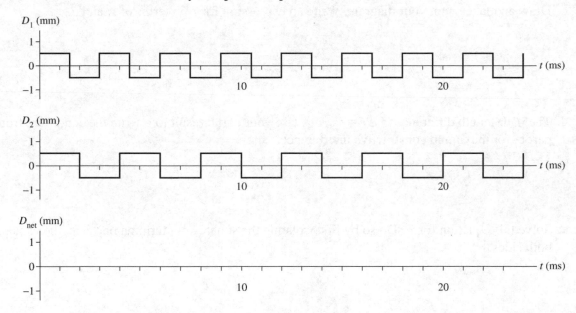

a. Period of wave 1? _____   Frequency of wave 1? _____

b. Period of wave 2? _____   Frequency of wave 2? _____

c. Draw the graph of the net wave at this point on the third set of axes. Be accurate, use a ruler!

d. Period of the net wave? _____   Frequency of the net wave? _____

e. Is the frequency of the superposition what you would expect as a beat frequency? Explain.

18 A Macroscopic Description of Matter

18.1 Solids, Liquids, and Gases

1. A 1×10^{-3} m^3 chunk of material has a mass of 3 kg.

   a. What is the material's density?

   b. Would a 2×10^{-3} m^3 chunk of the same material have the same mass? Explain.

   c. Would a 2×10^{-3} m^3 chunk of the same material have the same density? Explain.

2. You are given an irregularly shaped chunk of material and asked to find its density. List the *specific* steps that you would follow to do so.

3. Object 1 has an irregular shape. Its density is 4000 kg/m^3.

   a. Object 2 has the same shape and dimensions as object 1, but it is twice as massive. What is the density of object 2?

   b. Object 3 has the same mass and the same *shape* as object 1, but its size in all three dimensions is twice that of object 1. What is the density of object 3?

18.2 Atoms and Moles

4. You have 10 g of neon gas (^{20}Ne) and 10 g of nitrogen gas (^{14}N$_2$).

   a. Which gas consists of a larger number of moles? Explain.

   b. Which gas has a larger number of basic particles? Explain.

   c. Which gas has a larger number of *atoms*? Explain.

5. A cylinder contains 2 g of oxygen gas. A piston is used to compress the gas. After the gas has been compressed:

   a. Has the mass of the gas increased, decreased, or not changed? Explain.

   b. Has the density of the gas increased, decreased, or not changed? Explain.

   c. Have the number of moles of gas increased, decreased, or not changed? Explain.

   d. Has the number density of the gas increased, decreased, or not changed? Explain.

18.3 Temperature

6. Rank in order, from highest to lowest, the temperatures $T_1 = 0$ K, $T_2 = 0°C$, and $T_3 = 0°F$.

7. "Room temperature" is often considered to be 68°F. What is room temperature in °C and in K?

8. a. The gas pressure inside a sealed, rigid container is 1 atm at 100 K. What is the pressure at 200 K?

   b. The gas pressure inside a sealed, rigid container is 1 atm at 100°C. What is the pressure at 200°C?

18.4 Thermal Expansion

9. A strip of brass and a strip of steel are welded together. Brass has a larger coefficient of linear expansion than steel. The bimetal strip is straight at room temperature, as shown. What happens to the strip when the temperature increases? Be as specific as you can.

10. Blocks A, B, and C are made of the same metal and are at the same temperature. The dimensions shown are in centimeters.

    a. Rank in order, from most to least, the amount by which their heights increase if the temperature is increased. Explain.

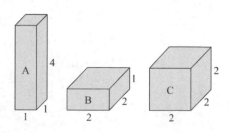

   Order:

   Explanation:

   b. Rank in order, from most to least, the amount by which their volumes increase if the temperature is increased. Explain.

   Order:

   Explanation:

18.5 Phase Changes

11. On the phase diagram:

    a. Draw *and label* a line to show a process in which the substance boils at constant pressure. Be sure to include an arrowhead on your line to show the direction of the process.

    b. Draw *and label* a line to show a process in which the substance freezes at constant temperature.

    c. Draw *and label* a line to show a process in which the substance sublimates at constant pressure.

    d. Draw a small circle around the critical point.

    e. Draw a small box around the triple point.

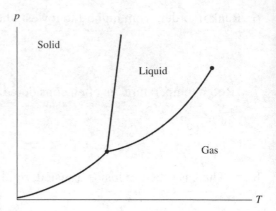

12. The figure shows the phase diagram of water. Answer the following questions by explicitly referring to the phase diagram and, perhaps, drawing lines on the phase diagram.

    a. What happens to the boiling-point temperature of water as you go to higher and higher elevations in the mountains?

    b. Suppose you place a beaker of liquid water at 20°C in a vacuum chamber and then steadily reduce the pressure. What, if anything, happens to the water?

    c. Is ice less dense or more dense than liquid water? Refer to the diagram as you answer.

18.6 Ideal Gases

13. It is well known that you can trap liquid in a drinking straw by placing the tip of your finger over the top while the straw is in the liquid, then lifting it out. The liquid runs out when you release your finger.

    a. What is the *net* force on the cylinder of trapped liquid?

    b. Draw a free-body diagram for the trapped liquid. Label each vector.

    c. Is the gas pressure inside the straw, between the liquid and your finger, greater than, less than, or equal to atmospheric pressure? Explain.

    d. If your answer to part c was "greater" or "less," how did the pressure change from the atmospheric pressure to the final pressure?

14. A gas is in a sealed container. By what factor does the gas pressure change if:

    a. The volume is tripled and the absolute temperature is doubled?

    b. The volume is halved and the absolute temperature is doubled?

15. A gas is in a sealed container. By what factor does the absolute gas temperature change if:

    a. The volume is tripled and the pressure is doubled?

    b. The volume is halved and the pressure is doubled?

16. The gas inside in a cylinder is heated, causing a piston in the cylinder to move outward. The heating causes the temperature to double and the length of the cylinder to triple. By what factor does the gas pressure change?

17. You have a 100 cm³ box of helium and a 100 cm³ box of argon.

    a. Suppose both boxes are at the same temperature and contain the same number of atoms. Is the helium pressure greater than, less than, or the same as the argon pressure? Explain.

    b. Suppose both gases have the same pressure and the same mass. Is the helium temperature greater than, less than, or the same as the argon temperature? Explain

18.7 Ideal-Gas Processes

18. The graphs below show the initial state of a gas. Draw a *pV* diagram showing the following processes:

   a. An isochoric process that doubles the pressure.

   b. An isobaric process that doubles the temperature.

   c. An isothermal process that doubles the volume.

19. Interpret the *pV* diagrams shown below by

   a. Naming the process.

   b. Stating the *factors* by which *p*, *V*, and *T* change. (A fixed quantity changes by a factor of 1.)

Process _____   Process _____   Process _____

p changes by _____   *p* changes by _____   *p* changes by _____

V changes by _____   *V* changes by _____   *V* changes by _____

T changes by _____   *T* changes by _____   *T* changes by _____

20. Starting from the initial state shown, draw a *pV* diagram for the three-step process:

   i. An isochoric process that halves the temperature, then

   ii. An isothermal process that halves the pressure, then

   iii. An isobaric process that doubles the volume.

   Label each of the stages on your diagram.

21. A cylinder of gas has a tightly fitting piston. If the piston is pushed in,
    i.   the pressure of the gas increases.
    ii. the pressure of the gas decreases.
    iii. the pressure of the gas remains the same.
    iv. we can't predict whether the pressure will increase, decrease, or remain the same.
    Which of these statements is true? Why?

22. A cylinder of gas has a tightly fitting piston. If the piston is pushed in,
    i.   the temperature of the gas increases.
    ii. the temperature of the gas decreases.
    iii. the temperature of the gas remains the same.
    iv. we can't predict whether the temperature will increase, decrease, or remain the same.
    Which of these statements is true? Why?

19 Work, Heat, and the First Law of Thermodynamics

19.1 It's All About Energy

19.2 Work in Ideal-Gas Processes

1. How much work is done on the gas in each of the following processes?

a.

b.

c.

W = _____   W = _____   W = _____

2. The figure on the left shows a process in which a gas is compressed from 300 cm³ to 100 cm³. On the right set of axes, draw the pV diagram of a process that starts from initial state i, compresses the gas to 100 cm³, and does the same amount of work on the gas as the process shown on the left.

3. The figure shows a process in which work is done to compress a gas.

   a. Draw and label a process A that starts and ends at the same points but does *more* work on the gas.

   b. Draw and label a process B that starts and ends at the same points but does *less* work on the gas.

4. An ideal gas undergoes a process described by $p = p_0 V_0^2/V^2$. At volume $V = V_0$, the pressure is p_0.
PSS How much work is done on the gas to compress it from $4V_0$ to V_0?
19.1
   a. One point on the pV diagram is shown, with pressure p_0.

   What is the pressure when $V = 2V_0$? $p =$ _____

   What is the pressure when $V = 4V_0$? $p =$ _____

   b. Based on your answers to part a, add two more points to
      the pV diagram, then sketch the curve of this ideal-gas
      process.

   c. In compressing the gas from $4V_0$ to V_0, is the work W positive
      or negative? Explain.

   d. On the pV diagram, shade the area that corresponds to the work W.

   e. Write the general expression, from PSS 19.1, for the work done on a gas.

   f. The p in your expression has to describe the gas process; it is a function of V. You know how
      pressure varies with volume for this process. You also know that the gas is being compressed from
      $4V_0$ to V_0. Make these substitutions.

   g. From calculus, you know that $\int_a^b f(x)\, dx = -\int_b^a f(x)\, dx$. Make this simplification.

   h. Finally, carry out the integration to find an expression for the work done.

19.3 Heat

19.4 The First Law of Thermodynamics

5. Metal blocks A and B, both at room temperature, are balanced on a see-saw. Block B is removed, heated to just below its melting point, then returned to the exact same position on the see-saw. Afterward, is block A higher than, lower than, or still level with block B? Explain.

6. Cold water is poured into a hot metal container.

   a. What physical quantities can you *measure* that tell you that the metal and water are somehow changing?

   b. What is the condition for equilibrium, after which no additional changes take place?

   c. Use the concept of energy to describe how the metal and the water interact.

   d. Is your description in part c something that you can *observe* happening? Or is it an *inference* based on the measurements you specified in part a?

7. Do each of the following describe a property of a system, an interaction of a system with its environment, or both? Explain.

   a. Temperature:

   b. Heat:

   c. Thermal energy:

8. Consider each of the following processes.

   a. Does the temperature increase (+), decrease (−), or not change (0)? Are the work W and the heat Q positive (+), negative (−), or zero (0)? Does the thermal energy increase (+), decrease (−), or not change (0)? Answer these questions by filling in the table.

	ΔT	W	Q	ΔE_{th}
You drive a nail into a board with a hammer.				
You hold a nail over a Bunsen burner.				
You compress the air in a bicycle pump by pushing down on the handle very rapidly.				
You turn on a flame under a cylinder of gas, and the gas undergoes an isothermal expansion.				
A flame turns liquid water into steam.				
High-pressure steam spins a turbine.				
A moving crate slides to a halt on a rough surface.				
High-pressure gas in a cylinder pushes a piston outward very rapidly.				

   b. Are each of your responses for W, Q, and ΔE_{th} consistent with the first law of thermodynamics? If not, which ones are not?

9. Consider an ideal-gas process that increases the volume of the gas in a cylinder without changing its pressure.

   a. Show the process on a pV diagram.

   b. Show the process on a first-law bar chart.

   c. What kind of process is this? _____

10. Consider an ideal-gas process that increases the pressure of the gas in a cylinder without changing its temperature.

   a. Show the process on a pV diagram.

   b. Show the process on a first-law bar chart.

   c. What kind of process is this? _____

11. The pV diagram shows two processes going from i to f.

   a. Is more work done on the gas in process A or in process B?

   Or is W the same for both? Explain. _____

   b. Is more heat energy transferred to the environment in process A or in process B? Or is Q the same for both? Explain. _____

19.5 Thermal Properties of Matter
19.6 Calorimetry

12. You have two 100 g cubes A and B, made of different materials. Cube A has a larger specific heat than cube B. Cube A, initially at 0°C, is placed in good thermal contact with cube B, initially at 200°C. The cubes are inside a well-insulated container where they don't interact with their surroundings. Is their final temperature greater than, less than, or equal to 100°C. Explain.

13. A beaker of water at 80.0°C is placed in the center of a well-insulated room whose air temperature is 20.0°C. Is the final temperature of the water:

    i.   20.0°C.
    ii. Slightly above 20.0°C.
    iii. 50.0°C.
    iv. Slightly below 80.0°C.
    v.   80.0°C.

    Explain.

14. 100 g of ice at 0°C and 100 g of steam at 100°C interact thermally in a well-insulated container. Is the final state of the system

    i.   An ice-water mixture at 0°C?
    ii. Water at a temperature between 0°C and 50°C?
    iii. Water at 50°C?
    iv. Water at a temperature between 50°C and 100°C?
    v.   A water-steam mixture at 100°C?

    Explain.

15. Mass M_A of metal A is molten, at a temperature T_A above the melting point T_m. It is poured
PSS into a mold made of a material B with a much higher melting temperature. The mold has
19.2 mass M_B and is initially at a temperature T_B low enough that metal A will solidify. What is their final
temperature T_f after the metal solidifies in the mold? The mold's specific heat is c_B. The metal has
specific heat c_A in the solid phase, c_A^* in the liquid phase, and heat of fusion L_A.

a. Does heat energy enter or leave system A? _____ System B? _____

b. Write an expression for Q_B, the heat required for B to reach the final temperature.

c. Heat Q_A consists of how many distinct terms? _____ Describe them.

d. Is each term in Q_A positive or negative? _____

e. Write an expression for Q_A, the heat required for A to reach the final temperature.

f. Knowing that $T_f < T_m$ and $T_m < T_A$, does each of the terms in your expression for Q_A have the sign
you claimed in part d? If not, rewrite your expression.

g. The calorimetry equation $Q_{net} = 0$ is based on the idea that the systems exchange energy but no heat
energy is transferred to or from the environment. Write the calorimetry equation for this process.

h. Finally, solve it for T_f.

19.7 The Specific Heats of Gases

16. You need to raise the temperature of a gas by 10°C. To use the least amount of heat energy, should you heat the gas at constant pressure or at constant volume? Explain.

17. The figure shows an adiabatic process.

    a. Is the final temperature higher than, lower than, or equal to the initial temperature?

    b. Draw *and label* the T_i and T_f isotherms on the figure.

    c. Is the work done on the gas positive or negative? Explain.

    d. Show *on the figure* how you would determine the amount of work done.

    e. Is any heat energy added to or removed from the system in this process? Explain.

    f. *Why* does the gas temperature change?

19.8 Heat-Transfer Mechanisms

18. A titanium cube at 400 K radiates 50 W of heat. How much heat does the cube radiate if its temperature is increased to 800 K?

19. Sphere A is 10 cm in diameter and at a temperature of 300 K. Sphere B, of the same material, is 5 cm in diameter and at a temperature of 600 K. Which sphere radiates more heat? Explain.

20 The Micro/Macro Connection

20.1 Molecular Speeds and Collisions

1. Solids and liquids resist being compressed. They are not totally incompressible, but it takes large forces to compress them even slightly. If it is true that matter consists of atoms, what can you infer about the microscopic nature of solids and liquids from their incompressibility?

2. a. Gases, in contrast with solids and liquids, are very compressible. What can you infer from this observation about the microscopic nature of gases?

   b. The density of air at STP is about $\frac{1}{1000}$ the density of water. How does the average distance between air molecules compare to the average distance between water molecules? Explain.

3. Can you think of any everyday experiences or observations that would suggest that the molecules of a gas are in constant, *random* motion? (Note: The existence of "wind" is *not* such an observation. Wind implies that the gas as a whole can move, but it doesn't tell you anything about the motions of the individual molecules in the gas.)

4. Helium has atomic mass number $A = 4$. Neon has $A = 20$ and argon has $A = 40$. Rank in order, from largest to smallest, the mean free paths λ_{He}, λ_{Ne}, and λ_{Ar} at STP. Explain.

20.2 Pressure in a Gas

5. According to kinetic theory, the pressure of a gas depends on the number density and the rms speed of the gas molecules. Consider a sealed container of gas that is heated at constant volume.

a. Does the number density of the gas increase or stay the same? Explain.

b. According to the ideal gas law, does the pressure of the gas increase or stay the same? Explain.

c. What can you infer from these observations about a relationship between the gas temperature (a macroscopic parameter) and the rms speed of the molecules (a microscopic parameter)?

6. Suppose you could suddenly increase the speed of every molecule in a gas by a factor of 2.

a. Would the rms speed of the molecules increase by a factor of $(2)^{1/2}$, 2, or 2^2? Explain.

b. Would the gas pressure increase by a factor of $(2)^{1/2}$, 2, or 2^2? Explain.

20.3 Temperature

7. If you double the absolute temperature of a gas:

    a. Does the average translational kinetic energy per molecule change? If so, by what factor?

    b. Does the root-mean-square velocity of the molecules change? If so, by what factor?

8. Lithium vapor, which is produced by heating lithium to the relatively low boiling point of 1340°C, forms a gas of Li_2 molecules. Each molecule has a molecular mass of 14 u. The molecules in nitrogen gas (N_2) have a molecular mass of 28 u. If the Li_2 and N_2 gases are at the same temperature, which of the following is true?

    i. v_{rms} of $N_2 = 2.00 \times v_{rms}$ of Li_2.
    ii. v_{rms} of $N_2 = 1.41 \times v_{rms}$ of Li_2.
    iii. v_{rms} of $N_2 = v_{rms}$ of Li_2.
    iv. v_{rms} of $N_2 = 0.71 \times v_{rms}$ of Li_2.
    v. v_{rms} of $N_2 = 0.50 \times v_{rms}$ of Li_2.

    Explain.

9. Suppose you could suddenly increase the speed of every molecule in a gas by a factor of 2. Would the absolute temperature of the gas increase by a factor of $(2)^{1/2}$, 2, or 2^2? Explain.

10. Two gases have the same number density and the same distribution of speeds. The molecules of gas 2 are more massive than the molecules of gas 1.

    a. Do the two gases have the same pressure? If not, which is larger?

    b. Do the two gases have the same temperature? If not, which is larger?

11. a. What is the average translational kinetic energy of a gas at absolute zero?

    b. Can a molecule have negative translational kinetic energy? Explain.

    c. Based on your answers to parts a and b, what is the translational kinetic energy of *every* molecule in the gas?

    d. Would it be physically possible for the thermal energy of a gas to be less than its thermal energy at absolute zero? Explain.

    e. Is it possible to have a temperature less than absolute zero? Explain.

20.4 Thermal Energy and Specific Heat

20.5 Thermal Interactions and Heat

12. Suppose you could suddenly increase the speed of every molecule in a gas by a factor of 2.
    a. Does the thermal energy of the gas change? If so, by what factor? If not, why not?

    b. Does the molar specific heat change? If so, by what factor? If not, why not?

13. Hot water is poured into a cold container. Give a *microscopic* description of how these two systems interact until they reach thermal equilibrium.

    Hot water

    Cold container

14. A beaker of cold water is placed over a flame.
    a. What *is* a flame?

    Water

    Flame

    b. Give a *microscopic* description of how the flame increases the water temperature.

15. The *rapid* compression of a gas by a fast-moving piston increases the gas temperature. For example, you likely have noticed that pumping up a bicycle tire causes the bottom of the pump to get warm. Consider a gas that is rapidly compressed by a piston.

    a. Does the thermal energy of the gas increase or stay the same? Explain.

    b. Is there a transfer of heat energy to the gas? Explain.

    c. Is work done on the gas? Explain.

    d. Give a *microscopic* description of why the gas temperature increases as the piston moves in.

16. A container with 0.1 mol of helium ($A = 4$), initially at 200°C, and a container with 0.1 mol of argon ($A = 40$), initially at 0°C, are placed in good thermal contact with each other. *After* they have reached thermal equilibrium:

    | 0.1 mol He | 0.2 mol Ar |

    a. Is v_{rms} of helium greater than, less than, or equal to v_{rms} of argon? Explain.

    b. Does the helium have more thermal energy than, less thermal energy than, or the same amount of thermal energy as the argon? Explain.

20.6 Irreversible Processes and the Second Law of Thermodynamics

17. Every cubic meter of air contains $\approx 200,000$ J of thermal energy. This is approximately the kinetic energy of a car going 40 mph. Even though it might be difficult to do, could a clever engineer design a car that uses the thermal energy already in the air as "fuel"? Even if only 1% of the thermal energy could be "extracted" from the air, it would take only ≈ 100 m^3 of air—the volume of a typical living room in a house—to get the car up to speed. Is this idea possible? Or does it violate the laws of physics?

18. If you place a jar of perfume in the center of a room and remove the stopper, you will soon be able to smell the perfume throughout the room. If you wait long enough, will all the perfume molecules ever be back in the jar at the same time? Why or why not?

19. Suppose you place an ice cube in a cup of room-temperature water and then seal them in a well-insulated container. No energy can enter or leave the container.

   a. If you open the container an hour later, which do you expect to find: a cup of water, slightly cooler than room temperature, or a large ice cube and some 100°C steam?

   b. Finding a large ice cube and some 100°C steam would not violate the first law of thermodynamics. $W = 0$ J and $Q = 0$ J, because the container is sealed, and $\Delta E_{th} = 0$ J because the increase in thermal energy of the water molecules that have become steam is offset by the decrease in thermal energy of water molecules that have turned to ice. Energy is conserved, yet we never see a process like this. Why not?

21 Heat Engines and Refrigerators

21.1 Turning Heat into Work

21.2 Heat Engines and Refrigerators

1. The figure on the left shows a thermodynamic process in which a gas expands from 100 cm³ to 300 cm³. On the right, draw the *pV* diagram of a process that starts from state i, expands to 300 cm³, and does the same amount of work as the process on the left.

2. For each of these processes, is work done *by* the system ($W < 0$, $W_s > 0$), *on* the system ($W > 0$, $W_s < 0$), or is *no* work done?

a.

Work is _____

b.

Work is _____

3. Rank in order, from largest to smallest, the thermal efficiencies η_1 to η_4 of these heat engines.

Order:
Explanation:

4. Could you have a heat engine with $\eta > 1$? Explain.

5. For each engine shown,
   a. Supply the missing value.
   b. Determine the thermal efficiency.

$\eta =$ _____

$\eta =$ _____

$\eta =$ _____

6. For each refrigerator shown,
   a. Supply the missing value.
   b. Determine the coefficient of performance.

$K =$ _____

$K =$ _____

$K =$ _____

7. Does a refrigerator do work in order to cool the interior? Explain.

21.3 Ideal-Gas Heat Engines

8. Starting from the point shown, draw a *pV* diagram for the following processes.

a. An isobaric process in which work is
   done *by* the system.

b. An adiabatic process in which work
   is done *on* the system.

c. An isothermal process in which heat is
   added to the system.

d. An isochoric process in which heat is
   removed from the system.

9. Rank in order, from largest to smallest, the amount of work $(W_s)_1$ to $(W_s)_4$ done by the gas in each of these cycles.

   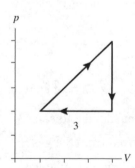

Order:

Explanation:

10. The figure uses a series of pictures to illustrate a thermodynamic cycle.

Stage 1 Stage 2 Pin Stage 3

The gas is compressed rapidly from V_1 to V_2.

The gas is heated at constant temperature until the volume returns to V_1.

The flame is turned off and the piston is locked in place.

The gas cools until the initial pressure p_1 is restored.

a. Show the cycle as a pV diagram. Label the three stages.
b. What is the energy transformation during each stage of the process? (For example, a stage in which work energy is transformed into heat energy could be represented as $W \rightarrow Q$.)

Stage 1: _____

Stage 2: _____

Stage 3: _____

11. The figure shows the pV diagram of a heat engine.

a. During which stages is heat added to the gas? _____

b. During which is heat removed from the gas? _____

c. During which stages is work done on the gas? _____

d. During which is work done by the gas? _____

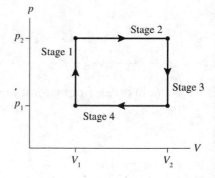

12. The figure shows the pV diagram of a heat engine.

a. During which stages is heat added to the gas? _____

b. During which is heat removed from the gas? _____

c. During which stages is work done on the gas? _____

d. During which is work done by the gas? _____

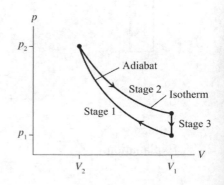

13. A heat engine satisfies $W_{out} = Q_{net}$. Why is there no ΔE_{th} term in this relationship?

14. The thermodynamic cycles of two heat engines are shown. Which engine has a larger thermal efficiency? Explain.

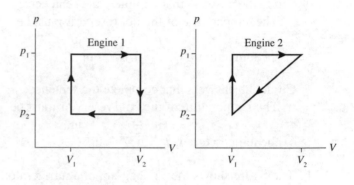

15. The thermodynamic cycles of two heat engines are shown. Which engine has a larger thermal efficiency? Explain.

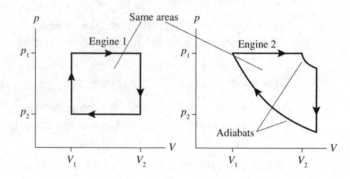

21.4 Ideal-Gas Refrigerators

16. a. The figure shows the pV diagram of a heat engine.

During which stages is heat added to the gas? _____

During which is heat removed from the gas? _____

During which stages is work done on the gas? _____

During which is work done by the gas? _____

Circle the answers that complete the sentence:
The temperature of the hot reservoir must be
    i. >   ii. =   iii. <
the temperature i. T_2 ii. T_3 iii. T_4

Circle the answers that complete the sentence:
The temperature of the cold reservoir must be
    i. >   ii. =   iii. <
the temperature i. T_2 ii. T_1 iii. T_4

b. The figure shows the pV diagram of a refrigerator.

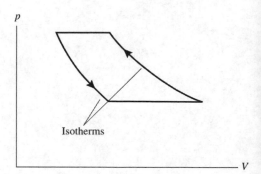

During which stages is heat added to the gas? _____

During which is heat removed from the gas? _____

During which stages is work done on the gas? _____

During which is work done by the gas? _____

Circle the answers that complete the sentence:
The temperature of the hot reservoir must be
    i. >   ii. =   iii. <
the temperature i. T_2 ii. T_3 iii. T_4

Circle the answers that complete the sentence:
The temperature of the cold reservoir must be
    i. >   ii. =   iii. <
the temperature i. T_2 ii. T_1 iii. T_4

17. An ideal-gas device operates with the cycle shown. Is it a refrigerator? That is, does it remove heat energy from a cold side and exhaust heat energy to a hot side? Explain.

21.5 The Limits of Efficiency

21.6 The Carnot Cycle

18. Do each of the following represent a possible heat engine or refrigerator? If not, what is wrong?

19. Four Carnot engines operate with the hot and cold reservoir temperatures shown in the table.

Engine	$T_C(K)$	$T_H(K)$
1	300	600
2	200	400
3	200	600
4	300	400

Rank in order, from largest to smallest, the thermal efficiencies η_1 to η_4 of these engines.

Order:

Explanation:

20. It gets pretty hot in your unairconditioned apartment. While browsing the Internet, you find a company selling small "room air conditioners." You place it on the floor, plug it in, and—the advertisement says—the air conditioner will lower the room temperature up to 10°F. Should you order one? Explain.

21. The first and second laws of thermodynamics are sometimes stated as "You can't win" and "You can't even break even." Do these sayings accurately reflect the laws of thermodynamics as applied to heat engines? Why or why not?

22 Electric Charges and Forces

22.1 The Charge Model

Note: Your answers in Section 22.1 should make *no* mention of electrons or protons.

1. Can an insulator be charged? If so, how would you charge an insulator? If not, why not?

2. Can a conductor be charged? If so, how would you charge a conductor? If not, why not?

3. Lightweight balls A and B hang straight down when both are neutral. They are close enough together to interact, but not close enough to touch. Draw pictures showing how the balls hang if:

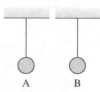

   a. Both are touched with a plastic rod that was rubbed with wool.

   b. The two charged balls of part a are moved farther apart.

c. Ball A is touched by a plastic rod that was rubbed with wool and ball B is touched by a glass rod that was rubbed with silk.

d. Both are charged by a plastic rod, but ball A is charged more than ball B.

e. Ball A is charged by a plastic rod. Ball B is neutral.

f. Ball A is charged by a glass rod. Ball B is neutral.

4. Four lightweight balls A, B, C, and D are suspended by threads. Ball A has been touched by a plastic rod that was rubbed with wool. When the balls are brought close together, without touching, the following observations are made:

   • Balls B, C, and D are attracted to ball A.
   • Balls B and D have no effect on each other.
   • Ball B is attracted to ball C.

   What are the charge states (glass, plastic, or neutral) of balls A, B, C, and D? Explain.

5. Charged plastic and glass rods hang by threads.

   a. An object repels the plastic rod. Can you predict what it will do to the glass rod? If so, what? If not, why not? Explain.

   b. A different object attracts the plastic rod. Can you predict what it will do to the glass rod? If so, what? If not, why not? Explain.

6. After combing your hair briskly, the comb will pick up small pieces of paper.

   a. Is the comb charged? Explain.

   b. How can you be sure that it isn't the paper that is charged? Propose an experiment to test this.

   c. Is your hair charged after being combed? What evidence do you have for your answer?

7. When you take clothes out of the drier right after it stops, the clothes often stick to your hands and arms. Is your body charged? If so, how did it acquire a charge? If not, why does this happen?

8. You've been given a piece of material. Propose an experiment or a series of experiments to determine if the material is a conductor or an insulator. State clearly what the outcome of each experiment will be if the material is a conductor and if it is an insulator.

9. Suppose there exists a third type of charge in addition to the two types we've called glass and plastic. Call this third type X charge. What experiment or series of experiments would you use to test whether an object has X charge? State clearly how each possible outcome of the experiments is to be interpreted.

22.2 Charge

22.3 Insulators and Conductors

10. A positively charged electroscope has separated leaves.

    a. Suppose you bring a positively charged rod close to the top of the electroscope, but not touching. How will the leaves respond? Use both charge diagrams and words to explain.

    b. How will the leaves respond if you bring a negatively charged rod close to the top of the electroscope, but not touching? Use both charge diagrams and words to explain.

11. a. A negatively charged plastic rod touches a neutral piece of metal. What is the final charge state (positive, negative, or neutral) of the metal? Use both charge diagrams and words to explain how this charge state is achieved.

    b. A positively charged glass rod touches a neutral piece of metal. What is the final charge state of the metal? Use both charge diagrams and words to explain how this charge state is achieved.

12. A lightweight, positively charged ball and a neutral rod hang by threads. They are close but not touching. A positively charged glass rod touches the hanging rod on the end opposite the ball, then the rod is removed.

   a. Draw a picture of the final positions of the hanging rod and the ball if the rod is made of glass.

   b. Draw a picture of the final positions of the hanging rod and the ball if the rod is metal.

13. Two oppositely charged metal spheres A and B have equal quantities of charge. They are brought into contact with a neutral metal rod.

   a. What is the final charge state of each sphere and of the rod?

   b. Give a microscopic explanation, in terms of fundamental charges, of how these final states are reached. Use both charge diagrams and words.

14. Metal sphere A has 4 units of negative charge and metal sphere B has 2 units of positive charge. The two spheres are brought into contact. What is the final charge state of each sphere? Explain.

15. a. Metal sphere A is initially neutral. A positively charged rod is brought near, but not touching. Is A now positive, negative, or neutral? Explain.

    b. Metal spheres A and B are initially neutral and are touching. A positively charged rod is brought near A, but not touching. Is A now positive, negative, or neutral? Explain.

    c. Metal sphere A is initially neutral. It is connected by a metal wire to the ground. A positively charged rod is brought near, but not touching. Is A now positive, negative, or neutral? Explain.

16. A lightweight, positively charged ball and a neutral metal rod hang by threads. They are close but not touching. A positively charged rod is held close to, but not touching, the hanging rod on the end opposite the ball.

    a. Draw a picture of the final positions of the hanging rod and the ball. Explain your reasoning.

    b. Suppose the positively charged rod is replaced with a negatively charged rod. Draw a picture of the final positions of the hanging rod and the ball. Explain your reasoning.

17. A positively charged rod is held near, but not touching, a neutral metal sphere.

    a. Add plusses and minuses to the figure to show the charge distribution on the sphere.

    b. Does the sphere experience a net force? If so, in which direction? Explain.

    Metal

18. If you bring your finger near a lightweight, negatively charged hanging ball, the ball swings over toward your finger. Use charge diagrams and words to explain this observation.

    Finger

19. The figure shows an atom with four protons in the nucleus and four electrons in the electron cloud.

    a. Draw a picture showing how this atom will look if a positive charge is held just *above* the atom.

    b. Is there a net force on the atom? If so, in which direction? Explain.

22.4 Coulomb's Law

20. For each pair of charged particles, draw a force vector on each particle to show the electric force on that particle. The length of each vector should be proportional to the magnitude of the force. Each + and − symbol represents the same quantity of charge.

    a.   (−)   (−)

    b.   (+ +)   (−)

    c.   (+ + / + +)   (−)

    d.   (+ + / + +)   (+ +)

21. For each group of charged particles, use a **black** pen or pencil to draw the forces acting on the gray charge. Then use a **red** pen or pencil to show the net force on the gray charge. Label $\vec{F}_{net}$.

    a. (+)   b. (−)   c.

                                                  (+ +)   (−)

         (+)   (+)

     (+)   (+)   (−)

22. Can you assign charges (positive or negative) so that these forces are correct? If so, show the charges on the figure. (There may be more than one correct response.) If not, why not?

    a.   b.

    c.   d.

23. Draw a + on the figure below at the position or positions where a proton would experience no net force.

    | | | (⁺⁺⁺) | | (+) | | | |

24. Draw a − on the figure below at the position or positions where an electron would experience no net force.

    | | | (⁺⁺⁺) | | (−) | | | |

25. The gray positive charge experiences a net force due to two other charges: the +1 charge that is seen and a +4 charge that is not seen. Add the +4 charge to the figure at the correct position.

26. Positive charges $4q$ and q are distance L apart. Let them be on
PSS the x-axis with $4q$ at the origin.
22.1

    a. Suppose a proton were placed on the x-axis to the *right* of q. Is it *possible* that the net electric force on the proton is zero? Explain.

    b. *On the figure*, draw a proton at an arbitrary point on the x-axis to the *left* of q, between $4q$ and q. Draw two force vectors and label them $\vec{F}_{4q}$ and $\vec{F}_q$ to show the two forces on this proton. Is it *possible* that, for the proper choice of r, the net electric force on the proton is zero? Explain.

    c. Write expressions for the magnitudes of forces $\vec{F}_{4q}$ and $\vec{F}_q$. Your expressions should be in terms of K, q, e, L, and r.

    $F_{4q} =$ _____ $F_q =$ _____

    d. Find the specific position—as a fraction of L—at which the net force is zero.

22.5 The Electric Field

27. At points 1 to 4, draw an electric field vector with the proper direction and whose length is proportional to the electric field strength at that point.

28. Dots 1 and 2 are two points in space. The electric field $\vec{E}_1$ at point 1 is shown. Can you determine the electric field at point 2? If so, draw it on the figure. If not, why not?

29. a. The electric field of a point charge is shown at *one* point in space.

900 N/C

   Can you tell if the point charge is + or −? If not, why not?

   b. Here the electric field of a point charge is shown at two positions in space.

900 N/C   400 N/C

   Now can you tell if the point charge is + or −? Explain.

   c. Can you determine the location of the charge? If so, draw it on the figure. If not, why not?

30. At the three points in space indicated with dots, draw the unit vector $\hat{r}$ that you would use to determine the electric field of the point charge.

    a.   b.

31. a. This is the unit vector $\hat{r}$ associated with a positive point charge. Draw the electric field vector at this point in space.

    b. This is the unit vector $\hat{r}$ associated with a negative point charge. Draw the electric field vector at this point in space.

32. The electric field strength at a point in space near a point charge is 1000 N/C.

    a. What will be the field strength if the quantity of charge is halved? Explain.

    b. What will be the field strength if the distance to the point charge is halved? The quantity of charge is the original amount, not the value of part a. Explain.

23 The Electric Field

23.1 Electric Field Models

23.2 The Electric Field of Point Charges

1. You've been assigned the task of determining the magnitude and direction of the electric field at a point in space. Give a step-by-step procedure of how you will do so. List any objects you will use, any measurements you will make, and any calculations you will need to perform. Make sure that your measurements do not disturb the charges that are creating the field.

2. Is there an electric field at the position of the dot? If so, draw the electric field vector on the figure. If not, what would you need to do to create an electric field at this point?

3. This is the electric field in a region of space.

    a. Explain the information that is portrayed in this diagram.

    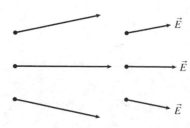

    b. If field vectors were drawn at the same six points but each was only half as long, would the picture represent the same electric field or a different electric field? Explain.

4. Each figure shows two vectors. Can a point charge create an electric field that looks like this at these two points? If so, draw the charge on the figure. If not, why not?

   Note: The dots are the points to which the vectors are attached. There are no charges at these points.

   a.   b.   c.   d.

5. At each of the dots, use a **black** pen or pencil to draw and label the electric fields $\vec{E}_1$ and $\vec{E}_2$ due to the two point charges. Make sure that the *relative* lengths of your vectors indicate the strength of each electric field. Then use a **red** pen or pencil to draw and label the net electric field $\vec{E}_{net}$ at each dot.

   a.   b.

   ⊕ q_1 •   ⊕ q_2 •   ⊖ q_1 •   ⊕ q_2 •

6. For each of the figures, use dots to mark any point or points (other than infinity) where $\vec{E} = \vec{0}$.

   a. | | | | ⊕ | | ⊕ | | | |

   b. | | | | (+ +/+ +) | | ⊖ | | | |

7. Compare the electric field strengths E_1 and E_2 at the two points labeled 1 and 2. For each, is $E_1 > E_2$, is $E_1 = E_2$ or is $E_1 < E_2$?

a.

(+) •1   (+) •2

_____

b.

(+) •1   (−) •2

_____

c.

•1 (+)   (−) •2

_____

d.

2•

(+) •1   (+)

_____

e.

(+) •1 (+/+) •2

_____

f.

(+) •1 (−/−) •2

_____

8. For each figure, draw and label the net electric field vector $\vec{E}_{net}$ at each of the points marked with a dot or, if appropriate, label the dot $\vec{E}_{net} = \vec{0}$. The lengths of your vectors should indicate the magnitude of $\vec{E}$ at each point.

a.

b.

c.

9. At the position of the dot, draw field vectors $\vec{E}_1$ and $\vec{E}_2$ due to q_1 and q_2, and the net electric field $\vec{E}_{net}$. Then, in the blanks, state whether the x- and y-components of $\vec{E}_{net}$ are positive or negative.

a.

$(E_{net})_x$ _____

$(E_{net})_y$ _____

b.

$(E_{net})_x$ _____

$(E_{net})_y$ _____

10. Draw the net electric field vector at the three points marked with a dot.

    Hint: Think of the charges as horizontal positive/negative pairs, then use superposition.

11. The figure shows the electric field lines in a region of space. Draw the electric field vectors at the three dots. The length of the vector should indicate the relative strength of the electric field at that point.

12. The figure shows the electric field lines in a region of space. Rank in order, from largest to smallest, the electric field strengths E_1 to E_4 at points 1 to 4.

    Order:

    Explanation:

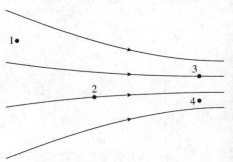

23.3 The Electric Field of a Continuous Charge Distribution

13. A small segment of wire contains 10 nC of charge.

    a. The segment is shrunk to one-third of its original length. What is the ratio λ_f/λ_i, where λ_i and λ_f are the initial and final linear charge densities?

    b. Suppose the original segment of wire is stretched to 10 times its original length. How much charge must be added to the wire to keep the linear charge density unchanged?

14. The figure shows two charged rods bent into a semicircle. For each, draw the electric field vector at the dot at the "center" of the semicircle.

    a.   b.

15. A hollow soda straw is uniformly charged. What is the electric field at the center (inside) of the straw? Explain.

16. A thin rod of length L is uniformly charged with total charge Q. What is the electric field strength on the axis of the rod at distance d from its center?

    PSS
    23.2
    a. Begin with a visual representation. Draw a horizontal rod, then divide it into 10 or 12 boxes with a $+$ in each box. Add an x-axis with the rod centered at the origin. Label the ends of the rod $x = -L/2$ and $x = L/2$. Put a dot on the x-axis at some point to the right of the rod; label it $x = d$.

    b. Pick one of your $+$ boxes to the right of the origin; label it "segment i," label its position as x_i, and write ΔQ beside it to show the charge in segment i. At the dot, draw the electric field vector due to segment i; label it $\vec{E}_i$.

c. Does $\vec{E}_i$ have an x-component? _____ A y-component? _____ A z-component? _____

d. Imagine the electric field $\vec{E}_j$ due to some other segment j.

Is $\vec{E}_j$ the same length as $\vec{E}_i$? _____ Does $\vec{E}_j$ point the same direction as $\vec{E}_i$? _____

e. The rod's electric field $\vec{E}$ is the sum of all the $\vec{E}_i$. Based on what you've said so far:

Does $\vec{E}$ have an x-component? _____ A y-component? _____ A z-component? _____

You should have found that $\vec{E}$ has only one component, requiring only one summation.

f. Using what you know about the electric field of a point charge, write an expression for the electric field component of $\vec{E}_i$—the component you identified in part e—in terms of ΔQ, x_i, d, and various constants.

g. Now write an expression of this component of the rod's field $\vec{E}$ as a sum over all i of your answer to part f.

h. The rod has charge Q in length L. What is the linear charge density? $\lambda =$ _____

i. Segment i has width Δx. Based on λ and Δx, the segment has charge $\Delta Q =$ _____

j. Rewrite your answer to part g with this substitution for ΔQ.

k. Now you're ready to convert the sum to an integral. What are the integration limits?

Lower limit _____ Upper limit _____

l. Write your expression for the electric field as a definite integral. That means (a) Change Δx to dx, (b) drop the subscript from x_i because x is now a continuous variable, (c) show the integration limits, and (d) take all multiplicative constants outside the integration.

We're going to stop here. You've done the physics by figuring out what to integrate. Now it's "just" a calculus problem of carrying out the integration to get a final answer.

23.4 The Electric Fields of Rings, Disks, Planes, and Spheres

17. An irregularly-shaped area of charge has surface charge density η_i.
    Each dimension (x and y) of the area is reduced by a factor of 3.163.

    a. What is the ratio η_f/η_i, where η_f is the final surface charge density?

    b. Compare the final force on a electron very far away to the initial force on the same electron.

18. A circular disk has surface charge density 8 nC/cm². What will be the surface charge density if the radius of the disk is doubled?

19. Rank in order, from largest to smallest, the surface charge densities η_1 to η_4 of surfaces 1 to 4.

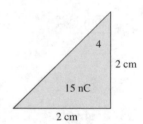

    Order:

    Explanation:

20. A sphere of radius R has charge Q. What happens to the electric field strength at $r = 2R$ if:
    a. The quantity of charge is halved?

    b. The radius of the sphere is halved?

23.5 The Parallel-Plate Capacitor

21. Rank in order, from largest to smallest, the electric field strengths E_1 to E_5 at each of these points.

Order:

Explanation:

22. A parallel-plate capacitor is constructed of two square plates, size $L \times L$, separated by distance d. The plates are given charge $\pm Q$. What is the ratio E_f/E_i of the final electric field strength E_f to the initial electric field strength E_i if:

   a. Q is doubled?

   b. L is doubled?

   c. d is doubled?

23. A ball hangs from a thread between two vertical capacitor plates. Initially, the ball hangs straight down. The capacitor plates are charged as shown, then the ball is given a small negative charge. The ball moves to one side, but not enough to touch a capacitor plate.

   a. Draw the ball and thread in the ball's new equilibrium position.

   b. In the space below, draw a free-body diagram of the ball when in its new position.

23.6 Motion of a Charged Particle in an Electric Field

23.7 Motion of a Dipole in an Electric Field

24. A small positive charge q experiences force $\vec{F}_1$ when placed at point 1. In terms of $\vec{F}_1$:

    a. What is the force on charge q at point 3? _____

    b. What is the force on a charge $3q$ at point 1? _____

    c. What is the force on a charge $2q$ at point 2? _____

    d. What is the force on a charge $-2q$ at point 2? _____

25. A small object is released from rest in the center of the capacitor. For each situation, does the object move to the right, to the left, or remain in place? If it moves, does it accelerate or move at constant speed?

    a. Positive object.

    b. Negative object.

    c. Neutral object.

26. Positively and negatively charged objects, with equal masses and equal quantities of charge, enter the capacitor in the directions shown.

    a. Use solid lines to draw their trajectories on the figure if their initial velocities are fast.

    b. Use dashed lines to draw their trajectories on the figure if their initial velocities are slow.

27. An electron is launched from the positive plate at a 45° angle. It does not have sufficient speed to make it to the negative plate. Draw its trajectory on the figure.

28. First a proton, later an electron are released from rest in the center of a capacitor.

    a. Compare the forces on the two charges. Are they equal, or is one larger? Explain.

    b. Compare the accelerations of the two charges. Are they equal, or is one larger? Explain.

29. The figure shows an electron orbiting a proton in a hydrogen atom.

    a. What force or forces act on the electron?

    b. Draw and label the following vectors on the figure: the electron's velocity $\vec{v}$ and acceleration $\vec{a}$, the net force $\vec{F}_{net}$ on the electron, and the electric field $\vec{E}$ at the position of the electron.

30. Does a charged particle always move in the direction of the electric field? If so, explain why. If not, give an example that is otherwise.

31. Three charges are placed at the corners of a triangle. The ++ charge has twice the quantity of charge of the two − charges; the net charge is zero.

    a. Draw the force vectors on each of the charges.

    b. Is the triangle in equilibrium? _____ If not, draw the equilibrium orientation directly beneath the triangle that is shown.

    c. Once in the equilibrium orientation, will the triangle move to the right, move to the left, rotate steadily, or be at rest? Explain.

24 Gauss's Law

24.1 Symmetry

1. An infinite plane of charge is seen edge on. The sign of the charge is not given. Do the electric fields shown below have the same symmetry as the charge? If not, why not?

a.

b.

c.

d.

2. Suppose you had a uniformly charged cube. Can you use symmetry alone to deduce the shape of the cube's electric field? If so, sketch and describe the field shape. If not, why not?

24.2 The Concept of Flux

3. The figures shown below are cross sections of three-dimensional closed surfaces. They have a flat top and bottom surface above and below the plane of the page. However, the electric field is everywhere parallel to the page, so there is no flux through the top or bottom surface. The electric field is uniform over each face of the surface. The field strength, in N/C, is shown.

   For each, does the surface enclose a net positive charge, a net negative charge, or no net charge?

a.

$Q_{in} =$ _____

b.

$Q_{in} =$ _____

c.

$Q_{in} =$ _____

d.

$Q_{in} =$ _____

e.

$Q_{in} =$ _____

f.

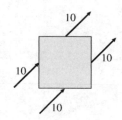

$Q_{in} =$ _____

4. The figures shown below are cross sections of three-dimensional closed surfaces. They have a flat top and bottom surface above and below the plane of the page, but there is no flux through the top or bottom surface. The electric field is perpendicular to and uniform over each face of the surface. The field strength, in N/C, is shown.

   Each surface contains no net charge. Draw the missing electric field vector (or write $\vec{E} = \vec{0}$) in the proper direction. Write the field strength beside it.

a.

b.

c.

d.

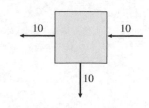

24.3 Calculating Electric Flux

5. Draw the area vector $\vec{A}$ for each of these surfaces.

6. How many area vectors are needed to characterize this closed surface? _____

   Draw them.

7. The diameter of the circle equals the edge length of the square. They are in a uniform electric field. Is the electric flux Φ_1 through the square larger than, smaller than, or equal to the electric flux Φ_2 through the circle? Explain.

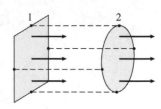

8. Is the electric flux Φ_1 through the circle larger than, smaller than, or equal to the electric flux Φ_2 through the hemisphere? Explain.

9. A uniform electric field is shown below.

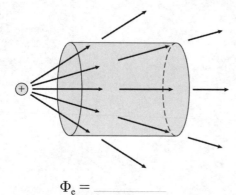

Draw and label an *edge view* of three square surfaces, all the *same size*, for which
a. The flux is maximum.
b. The flux is minimum.
c. The flux has half the value of the flux through the square part of a.

Give the tilt angle of any squares not perpendicular or parallel to the field lines.

10. Is the net electric flux through each of the closed surfaces below positive (+), negative (−), or zero (0)?

a.

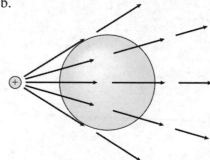

$\Phi_e = $ _____

b.

$\Phi_e = $ _____

c.

Charge inside

$\Phi_e = $ _____

d.

Charge inside

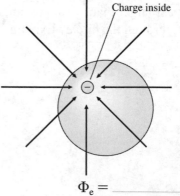

$\Phi_e = $ _____

e.

Charge inside

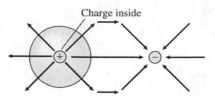

$\Phi_e = $ _____

f.

Charges inside

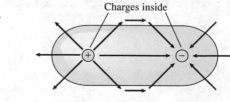

$\Phi_e = $ _____

24.4 Gauss's Law

24.5 Using Gauss's Law

11. For each of the closed cylinders shown below, are the electric fluxes through the top, the wall, and the bottom positive (+), negative (−), or zero (0)? Is the net flux positive, negative, or zero?

a.

$\Phi_{top} =$ _____
$\Phi_{wall} =$ _____
$\Phi_{bot} =$ _____
$\Phi_{net} =$ _____

b.

$\Phi_{top} =$ _____
$\Phi_{wall} =$ _____
$\Phi_{bot} =$ _____
$\Phi_{net} =$ _____

c.

$\Phi_{top} =$ _____
$\Phi_{wall} =$ _____
$\Phi_{bot} =$ _____
$\Phi_{net} =$ _____

d.

$\Phi_{top} =$ _____
$\Phi_{wall} =$ _____
$\Phi_{bot} =$ _____
$\Phi_{net} =$ _____

e.

$\Phi_{top} =$ _____
$\Phi_{wall} =$ _____
$\Phi_{bot} =$ _____
$\Phi_{net} =$ _____

f.

$\Phi_{top} =$ _____
$\Phi_{wall} =$ _____
$\Phi_{bot} =$ _____
$\Phi_{net} =$ _____

g.

$\Phi_{top} =$ _____
$\Phi_{wall} =$ _____
$\Phi_{bot} =$ _____
$\Phi_{net} =$ _____

h.

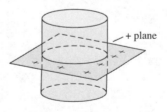

$\Phi_{top} =$ _____
$\Phi_{wall} =$ _____
$\Phi_{bot} =$ _____
$\Phi_{net} =$ _____

i.

$\Phi_{top} =$ _____
$\Phi_{wall} =$ _____
$\Phi_{bot} =$ _____
$\Phi_{net} =$ _____

12. For this closed cylinder, $\Phi_{\text{top}} = -15 \text{ Nm}^2/\text{C}$ and $\Phi_{\text{bot}} = 5 \text{ Nm}^2/\text{C}$. What is Φ_{wall}?

13. What is the electric flux through each of these closed surfaces? (The charges in the gray areas are inside the closed surfaces.) Give your answers as multiples of q/ϵ_0.

a.

b.

c.

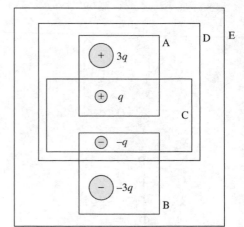

$\Phi_{\text{e}} =$ _____

$\Phi_{\text{e}} =$ _____

$\Phi_{\text{e}} =$ _____

14. What is the electric flux through each of these surfaces? Give your answers as multiples of q/ϵ_0.

$\Phi_{\text{A}} =$ _____

$\Phi_{\text{B}} =$ _____

$\Phi_{\text{C}} =$ _____

$\Phi_{\text{D}} =$ _____

$\Phi_{\text{E}} =$ _____

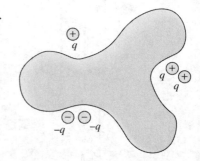

15. A positively charged balloon expands as it is blown up, increasing in size from the initial to final diameters shown. Do the electric fields at points 1, 2, and 3 increase, decrease, or stay the same? Explain your reasoning for each.

Point 1:

Point 2:

Point 3:

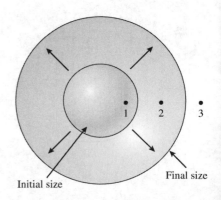

16. Three charges, all the same charge q, are surrounded by three spheres of equal radii.

a. Rank in order, from largest to smallest, the fluxes Φ_1, Φ_2, and Φ_3 through the spheres.

Order:

Explanation:

b. Rank in order, from largest to smallest, the electric field strengths E_1, E_2, and E_3 on the surfaces of the spheres.

Order:

Explanation:

17. Two spheres of different diameters surround equal charges. Three students are discussing the situation.

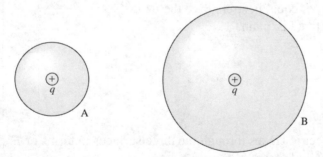

Student 1: The flux through spheres A and B are equal because they enclose equal charges.

Student 2: But the electric field on sphere B is weaker than the electric field on sphere A. The flux depends on the electric field strength, so the flux through A is larger than the flux through B.

Student 3: I thought we learned that flux was about surface area. Sphere B is larger than sphere A, so I think the flux through B is larger than the flux through A.

Which of these students, if any, do you agree with? Explain.

18. A sphere and an ellipsoid surround equal charges. Four students are discussing the situation.

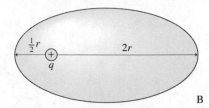

A

B

Student 1: The fluxes through A and B are equal because the average radius is the same.

Student 2: I agree that the fluxes are equal, but it's because they enclose equal charges.

Student 3: The electric field is not perpendicular to the surface for B, and that makes the flux through B less than the flux through A.

Student 4: I don't think that Gauss's law even applies to a situation like B, so we can't compare the fluxes through A and B.

Which of these students, if any, do you agree with? Explain.

19. Two parallel, infinite planes of charge have charge densities
PSS 2η and $-\eta$. A Gaussian cylinder with cross section area A extends
24.1 distance L to either side.

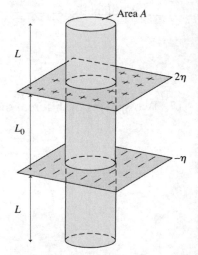

a. Is $\vec{E}$ perpendicular or parallel to the Gaussian surface at the:

Top _____ Bottom _____ Wall _____

b. Is the electric field E_{top} emerging from the top surface stronger than, weaker than, or equal in strength to the field E_{bot} emerging from the bottom? Explain.

c. By inspection, write the electric fluxes through the three surfaces in terms of E_{top}, E_{bot}, E_{wall}, L, L_0, and A. (You may not need all of these.)

$\Phi_{top} = $ _____ $\Phi_{bot} = $ _____ $\Phi_{wall} = $ _____

d. How much charge is enclosed within the cylinder? Write Q_{in} in terms of η, L, L_0, and A.

$Q_{in} = $ _____

e. By combining your answers from parts b, c, and d, use Gauss's law to determine the electric field strength above the top plane. Show your work.

24.6 Conductors in Electrostatic Equilibrium

20. A small metal sphere hangs by a thread within a larger, hollow conducting sphere. A charged rod is used to transfer positive charge to the outer surface of the hollow sphere.

   a. Suppose the thread is an insulator. After the charged rod touches the outer sphere and is removed, are the following surfaces positive, negative, or not charged?

   The small sphere: _____

   The inner surface of the hollow sphere: _____

   The outer surface of the hollow sphere: _____

   b. Suppose the thread is a conductor. After the charged rod touches the outer sphere and is removed, are the following surfaces positive, negative, or not charged?

   The small sphere: _____

   The inner surface of the hollow sphere: _____

   The outer surface of the hollow sphere: _____

21. A small metal sphere hangs by an insulating thread within a larger, hollow conducting sphere. A conducting wire extends from the small sphere through, but not touching, a small hole in the hollow sphere. A charged rod is used to transfer positive charge to the wire. After the charged rod has touched the wire and been removed, are the following surfaces positive, negative, or not charged?

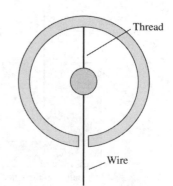

   The small sphere: _____

   The inner surface of the hollow sphere: _____

   The outer surface of the hollow sphere: _____

22. A −10 nC point charge is inside a hole in a conductor. The conductor has no net charge.

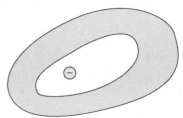

   a. What is the total charge on the inside surface of the conductor?

   b. What is the total charge on the outside surface of the conductor?

23. A −10 nC point charge is inside a hole in a conductor. The conductor has a net charge of +10 nC.

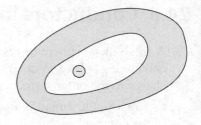

a. What is the total charge on the inside surface of the conductor?

b. What is the total charge on the outside surface of the conductor?

24. An insulating thread is used to lower a positively charged metal ball into a metal container. Initially, the container has no net charge. Use plus and minus signs to show the charge distribution on the ball and the container at the times shown in the figure. (The ball's charge is already shown in the first frame.)

Ball hasn't touched

Ball has touched

Ball has been withdrawn

25 The Electric Potential

25.1 Electric Potential Energy

25.2 The Potential Energy of Point Charges

1. A positive point charge and a negative point charge are inside a parallel-plate capacitor. The point charges interact only with the capacitor, not with each other. Let the negative capacitor plate be the zero of potential energy for both charges.

   a. Use a **black** pen or pencil to draw the electric field vectors inside the capacitor.

   b. Use a **red** pen or pencil to draw the forces acting on the two charges.

   c. Is the potential energy of the *positive* point charge positive, negative, or zero? Explain.

   d. In which direction (right, left, up, or down) does the potential energy of the positive charge decrease? Explain.

   e. In which direction will the positive charge move if released from rest? Use the concept of energy to explain your answer.

   f. Does your answer to part e agree with the force vector you drew in part b? _____

   g. Repeat steps c to f for the *negative* point charge.

2. A positive charge q is fired through a small hole in the positive plate of a capacitor. Does q speed up or slow down inside the capacitor? Answer this question twice:

   a. First using the concept of force.

   b. Second using the concept of energy.

3. Charge $q_1 = 3$ nC is distance r from a positive point charge Q. Charge $q_2 = 1$ nC is distance $2r$ from Q. What is the ratio U_1/U_2 of their potential energies due to their interactions with Q?

4. The figure shows the potential energy of a positively charged particle in a region of space.

   a. What possible arrangement of source charges is responsible for this potential energy? Draw the source charges above the axis below.

   b. With what kinetic energy should the charged particle be launched from $x = 0$ mm to have a turning point at $x = 3$ mm? Explain.

   c. How much kinetic energy does this charged particle of part b have as it passes $x = 2$ mm?

5. An electron ($q = -e$) completes half of a circular orbit of radius r around a nucleus with $Q = +3e$.

   a. How much work is done on the electron as it moves from i to f? Give either a numerical value or an expression from which you could calculate the value if you knew the radius. Justify your answer.

   b. By how much does the electric potential energy change as the electron moves from i to f?

   c. Is the electron's speed at f greater than, less than, or equal to its speed at i?

   d. Are your answers to parts a and c consistent with each other?

6. An electron moves along the trajectory from i to f.

   a. Does the electric potential energy increase, decrease, or stay the same? Explain.

   b. Is the electron's speed at f greater than, less than, or equal to its speed at i? Explain.

7. Inside a parallel-plate capacitor, two protons are launched with the same speed from point 1. One proton moves along the path from 1 to 2, the other from 1 to 3. Points 2 and 3 are the same distance from the negative plate.

a. Is $\Delta U_{1 \rightarrow 2}$, the change in potential energy along the path $1 \rightarrow 2$, larger than, smaller than, or equal to $\Delta U_{1 \rightarrow 3}$? Explain.

b. Is the proton's speed v_2 at point 2 larger than, smaller than, or equal to v_3? Explain.

25.3 The Potential Energy of a Dipole

8. Rank in order, from most positive to most negative, the potential energies U_1 to U_6 of these six electric dipoles in a uniform electric field.

Order:

Explanation:

25.4 The Electric Potential

9. Charged particles with $q = +0.1$ C are fired with 10 J of kinetic energy toward a region of space in which there is an electric potential. The figure shows the kinetic energy of the charged particles as they arrive at nine different points in the region. Determine the electric potential at each of these points. Write the value of the potential *above* each of the dots. Assume that the particles start from a point where the electric potential is zero.

10. a. The graph on the left shows the electric potential along the *x*-axis. Use the axes on the right to draw a graph of the potential energy of a 0.1 C charged particle in this region of space. Provide a numerical scale on the energy axis.

   b. If the charged particle is shot toward the right from $x = 1$ m with 1.0 J of kinetic energy, where is its turning point? Explain.

   c. Will the charged particle of part b ever reach $x = 0$ m? If so, how much kinetic energy will it have at that point? If not, why not?

25.5 The Electric Potential Inside a Parallel-Plate Capacitor

11. Rank in order, from largest to smallest, the electric potentials V_1 to V_5 at points 1 to 5.

   Order:

   Explanation:

12. The figure shows two points inside a capacitor. Let $V = 0$ V at the negative plate.

   a. What is the ratio V_2/V_1 of the electric potentials at these two points? Explain.

   b. What is the ratio E_2/E_1 of the electric field strengths at these two points? Explain.

13. The figure shows two capacitors, each with a 3 mm separation. A proton is released from rest in the center of each capacitor.

   a. Draw an arrow on each proton to show the direction it moves.

   b. Which proton reaches a capacitor plate first? Or are they simultaneous? Explain.

14. A capacitor with plates separated by distance d is charged to a potential difference ΔV_C. All wires and batteries are disconnected, then the two plates are pulled apart (with insulated handles) to a new separation of distance $2d$.

a. Does the capacitor charge Q change as the separation increases? If so, by what factor? If not, why not?

b. Does the electric field strength E change as the separation increases? If so, by what factor? If not, why not?

c. Does the potential difference ΔV_C change as the separation increases? If so, by what factor? If not, why not?

15. Each figure shows a contour map on the left and a set of graph axes on the right. Draw a graph of V versus x. Your graph should be a straight line or a smooth curve.

a.

b.

c.

16. Each figure shows a *V*-versus-*x* graph on the left and an *x*-axis on the right. Assume that the potential varies with *x* but not with *y*. Draw a contour map of the electric potential. Your figures should look similar to the contour maps in Question 15. Space your equipotential lines every 20 volts and label them.

a.

b.

25.6 The Electric Potential of a Point Charge

17. Rank in order, from largest to smallest, the electric potentials V_1 to V_5 at points 1 to 5.

   Order:

   Explanation:

18. Rank in order, from least negative to most negative, the electric potentials V_1 to V_5 at points 1 to 5.

   Order:

   Explanation:

19. The figure shows two points near a positive point charge.

   a. What is the ratio V_1/V_2 of the electric potentials at these two points? Explain.

   b. What is the ratio E_1/E_2 of the electric field strengths at these two points? Explain.

20. A 1 nC positive point charge is located at point A. The electric potential at point B is

    a. 9 V   b. $9 \cdot \sin 30°$ V   c. $9 \cdot \cos 30°$ V   d. $9 \cdot \tan 30°$ V

    Explain the reason for your choice.

21. An inflatable metal balloon of radius R is charged to a potential of 1000 V. After all wires and batteries are disconnected, the balloon is inflated to a new radius $2R$.

    a. Does the potential of the balloon change as it is inflated? If so, by what factor? If not, why not?

    b. Does the potential at a point at distance $r = 4R$ change as the balloon is inflated? If so, by what factor? If not, why not?

22. A small charged sphere of radius R_1, mass m_1, and positive charge q_1 is shot head on with speed v_1
PSS from a long distance away toward a second small sphere having radius R_2, mass m_2, and positive charge
25.1 q_2. The second sphere is held in a fixed location and cannot move. The spheres repel each other, so
sphere 1 will slow as it approaches sphere 2. If v_1 is small, sphere 1 will reach a closest point, reverse
direction, and be pushed away by sphere 2. If v_1 is large, sphere 1 will crash into sphere 2. For what
speed v_1 does sphere 1 just barely touch sphere 2 as it reverses direction?

a. Begin by drawing a before-and-after pictorial representation. Initially, the spheres are far apart and
   sphere 1 is heading toward sphere 2 with speed v_1. The problem ends with the spheres touching.
   What is speed of sphere 1 at this instant? How far apart are the centers of the spheres at this instant?
   Label the before and after pictures with complete information—all in symbolic form.

b. Energy is conserved, so we can use Problem-Solving Strategy 25.1. But first we have to identify the
   "moving charge" q and the "source charge" that creates the potential.

   Which is the moving charge? _____   Which is the source charge? _____

c. We're told the charges start "a long distance away" from each other. Based on this statement, what
   value can you assign to V_i, the potential of the source charge at the initial position of the moving
   charge? Explain.

d. Now write an expression in terms of the symbols defined above (and any constants that are needed)
   for the initial energy $K_i + qV_i$.

   $K_i + qV_i =$ _____

e. Referring to information on your visual overview, write an expression for the final energy.

   $K_f + qV_f =$ _____

f. Energy is conserved, so finish the problem by solving for v_1.

25.7 The Electric Potential of Many Charges

23. Each figure below shows three points in the vicinity of two point charges. The charges have equal magnitudes. Rank in order, from largest to smallest, the potentials V_1, V_2, and V_3.

a.

b.

c.

d.

24. On the axes below, draw a graph of V versus x for the two point charges shown.

a.

b.

b.

25. For each pair of charges below, are there any points (other than at infinity) at which the electric potential is zero? If so, show them on the figure with a dot and a $V = 0$ label. If not, why not?

a.

b.

26. For each pair of charges below, at which grid point or points could a double-negative point charge
    ($q = -2$) be placed so that the potential at the dot is 0 V? There may be more than one possible point.
    Draw the charge on the figure at all points that work.

    a.

    b.

27. The graph shows the electric potential
    along the x-axis due to point charges on the
    x-axis.

    a. Draw the charges *on the axis of the*
       figure. Note that the charges may have
       different magnitudes.

    b. An electron is placed at $x = 2$ cm. Is its
       potential energy positive, negative,
       or zero? Explain.

    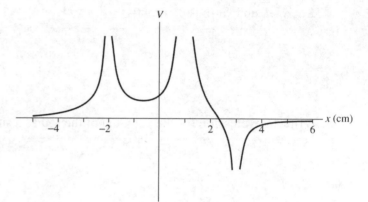

    c. If the electron is released from rest at $x = 2$ cm, will it move right, move left, or remain at $x = 2$ cm?
       Base your explanation on energy concepts.

28. A ring has radius R and charge Q. The ring is shrunk to a new radius $\frac{1}{2}R$ with no change in its charge.
    By what factor does the on-axis potential at $z = R$ increase?

29. A thin rod of length L is uniformly charged with total charge Q. What is the electric
PSS potential on the axis of the rod at distance d from its center?
25.2
   a. Begin with a visual representation. Draw a
      horizontal rod, then divide it into 10 or 12
      boxes with a $+$ in each box. Add an x-axis
      with the rod centered at the origin. Label the
      ends of the rod $x = -L/2$ and $x = L/2$. Put a
      dot on the x-axis at some point to the right of
      the rod; label it $x = d$.

   b. Pick one of your $+$ boxes to the right of the origin; label it "segment i," label its position as x_i, and
      write ΔQ beside it to show the charge in segment i.

   c. Using what you know about the electric potential of a point charge, write an expression for the
      electric potential of segment i at the position of the dot. Your expression should be in terms of
      ΔQ, x_i, d, and various constants.

   d. Write an expression for the potential of the rod as a sum over all i of your answer to part c.

   e. The rod has charge Q in length L. What is the linear charge density? $\lambda =$ _____

   f. Segment i has width Δx. Based on λ and Δx, the segment has charge $\Delta Q =$ _____

   g. Rewrite your answer to part d with this substitution for ΔQ.

   h. Now you're ready to convert the sum to an integral. What are the integration limits?

      Lower limit _____   Upper limit _____

   i. Write your expression for the potential as a definite integral. That means (a) Change Δx to dx, (b)
      drop the subscript from x_i because x is now a continuous variable, (c) show the integration limits,
      and (d) take all multiplicative constants outside the integration.

      We're going to stop here. You've done the physics by figuring out what to integrate. Now it's "just" a
      calculus problem of carrying out the integration to get a final answer.

26 Potential and Field

26.1 Connecting Potential and Field

1. The top graph shows the x-component of $\vec{E}$ as a function of x. On the axes below the graph, draw the graph of V versus x in this same region of space. Let $V = 0$ V at $x = 0$ m. Include an appropriate vertical scale. (Hint: Integration is the area under the curve.)

a.

b.

c.

d.

26.2 Finding the Electric Field from the Potential

2. The top graph shows the electric potential as a function of x. On the axes below the graph, draw the graph of E_x versus x in this same region of space. Add an appropriate scale on the vertical axis.

a.

b.

3. For each contour map:

   i. Estimate the electric fields $\vec{E}_a$ and $\vec{E}_b$ at points a and b. Don't forget that $\vec{E}$ is a vector. Show how you made your estimate.
   ii. On the contour map, draw the electric field vectors at points a and b.

a.

$\vec{E}_a -$ _____

$\vec{E}_b =$ _____

b.

$\vec{E}_a =$ _____

$\vec{E}_b =$ _____

4. The top graph shows E_x versus x for an electric field that is parallel to the x-axis.

   a. Draw the graph of V versus x in this region of space. Let $V = 0$ V at $x = 0$ m. Add an appropriate scale on the vertical axis. (Hint: Integration is the area under the curve.)

   b. Use dashed lines to draw a contour map above the x-axis on the right. Space your equipotential lines every 20 volts and label each equipotential line.

   c. Draw electric field vectors on top of the contour map.

5. Draw the electric field vectors at the dots on this contour map. The length of each vector should be proportional to the field strength at that point.

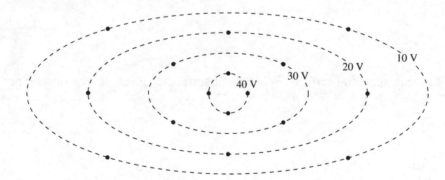

6. Draw the electric field vectors at the dots on this contour map. The length of each vector should be proportional to the field strength at that point.

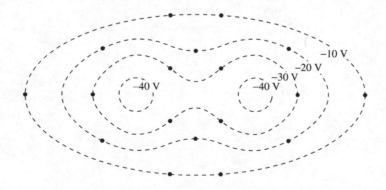

7. a. Suppose $\vec{E} = \vec{0}$ V/m throughout some region of space. Is $V = 0$ V in this region? Explain.

   b. Suppose $V = 0$ V throughout some region of space. Is $\vec{E} = \vec{0}$ V/m in this region? Explain.

8. The figure shows an electric field diagram. Dashed lines 1 and 2 are two surfaces in space, not physical objects.

   a. Is the electric potential at point a higher than, lower than, or equal to the electric potential at point b? Explain.

   b. Rank in order, from largest to smallest, the magnitudes of potential differences ΔV_{ab}, ΔV_{cd}, and ΔV_{ef}.

   Order:

   Explanation:

   c. Is surface 1 an equipotential surface? What about surface 2? Explain why or why not.

9. For each of the figures below, is this a physically possible potential map if there are no free charges in this region of space? If so, draw an electric field line diagram on top of the potential map. If not, why not?

   a.

   b.

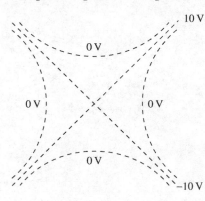

26.3 A Conductor in Electrostatic Equilibrium

10. The figure shows a negatively charged electroscope. The gold leaf stands away from the rigid metal post. Is the electric potential of the leaf higher than, lower than, or equal to the potential of the post? Explain.

11. Two metal spheres are connected by a metal wire that has a switch in the middle. Initially the switch is open. Sphere 1, with the larger radius, is given a positive charge. Sphere 2, with the smaller radius, is neutral. Then the switch is closed. Afterward, sphere 1 has charge Q_1, is at potential V_1, and the electric field strength at its surface is E_1. The values for sphere 2 are Q_2, V_2, and E_2.

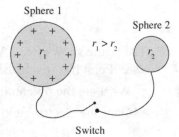

   a. Is V_1 larger than, smaller than, or equal to V_2? Explain.

   b. Is Q_1 larger than, smaller than, or equal to Q_2? Explain.

   c. Is E_1 larger than, smaller than, or equal to E_2? Explain.

12. The figure shows two flat metal electrodes that are held at potentials of 100 V and 0 V.
    a. Used dashed lines to sketch a reasonable approximation of the 20 V, 40 V, 60 V, and 80 V equipotential lines.
    b. Draw enough electric field lines to indicate the shape of the electric field.

13. The figure shows two 3 V batteries with metal wires attached to each end. Points a and c are *inside* the wire. Point b is inside the battery. For each figure:
    * What are the potential differences ΔV_{12}, ΔV_{23}, ΔV_{34}, and ΔV_{14}? Define $\Delta V_{ab} = V_a - V_b$.
    * Does the electric field at a, b, and c point left, right, up, or down? Or is $\vec{E} = \vec{0}$?

a.

b.

$\Delta V_{12} = $ _____ $\Delta V_{23} = $ _____

$\Delta V_{34} = $ _____ $\Delta V_{14} = $ _____

$\vec{E}_a$ _____ $\vec{E}_b$ _____ $\vec{E}_c$ _____

$\Delta V_{12} = $ _____ $\Delta V_{23} = $ _____

$\Delta V_{34} = $ _____ $\Delta V_{14} = $ _____

$\vec{E}_a$ _____ $\vec{E}_b$ _____ $\vec{E}_c$ _____

26.4 Sources of Electric Potential

14. What is ΔV_{series} for each group of 1.5 V batteries?

    a. | −1.5 V+ |−1.5 V+ |−1.5 V+ |−1.5 V+| $\Delta V_{series} = $ _____

    b. | −1.5 V+ || +1.5 V− |−1.5 V+ |−1.5 V+| $\Delta V_{series} = $ _____

    c. | −1.5 V+ || +1.5 V −|−1.5 V+ || +1.5 V −| $\Delta V_{series} = $ _____

26.5 Capacitance and Capacitors

26.6 The Energy Stored in a Capacitor

15. A parallel-plate capacitor with plate separation d is connected to a battery that has potential difference ΔV_{bat}. Without breaking any of the connections, insulating handles are used to increase the plate separation to $2d$.

    a. Does the potential difference ΔV_C change as the separation increases? If so, by what factor? If not, why not?

    b. Does the capacitance change? If so, by what factor? If not, why not?

    c. Does the capacitor charge Q change? If so, by what factor? If not, why not?

16. For the capacitor shown, the potential difference ΔV_{ab} between points a and b is

    a. 6 V   b. $6 \cdot \sin 30°$ V   c. $6/\sin 30°$ V

    d. $6 \cdot \tan 30°$ V   e. $6 \cdot \cos 30°$ V   f. $6/\cos 30°$ V

    Explain your choice.

17. Rank in order, from largest to smallest, the potential differences $(\Delta V_C)_1$ to $(\Delta V_C)_4$ of these four capacitors.

Order:

Explanation:

18. Each capacitor in the circuits below has capacitance C. What is the equivalent capacitance of the group of capacitors?

a.

$C_{eq} =$ _____

b.

$C_{eq} =$ _____

c.

$C_{eq} =$ _____

d.

$C_{eq} =$ _____

e.

$C_{eq} =$ _____

f.

$C_{eq} =$ _____

19. Rank in order, from largest to smallest, the equivalent capacitances $(C_{eq})_1$ to $(C_{eq})_4$ of these four groups of capacitors.

Order:

Explanation:

20. Rank in order, from largest to smallest, the energies $(U_C)_1$ to $(U_C)_4$ stored in each of these capacitors.

Order:

Explanation:

26.7 Dielectrics

21. An air-insulated capacitor is charged until the electric field strength inside is 10,000 V/m, then disconnected from the battery. When a dielectric is inserted between the capacitor plates, filling the space, the electric field strength is reduced to 2000 V/m.

    a. Does the amount of charge on the capacitor plates increase, decrease, or stay the same when the dielectric is inserted? If it increases or decreases, by what factor?

    b. Does the potential difference between the capacitor plates increase, decrease, or stay the same when the dielectric is inserted? If it increases or decreases, by what factor?

22. An air-insulated capacitor is charged until the plates have charge $\pm 10\ \mu C$, then left connected to the battery. When a dielectric is inserted between the capacitor plates, filling the space, the charge increases to $\pm 50\ \mu C$.

    a. Does the potential difference across the capacitor plates increase, decrease, or stay the same when the capacitor is inserted? If it increases or decreases, by what factor?

    b. Does the electric field between the capacitor plates increase, decrease, or stay the same when the capacitor is inserted? If it increases or decreases, by what factor?

23. The gap between two capacitor plates is *partially* filled with a dielectric. Rank in order, from largest to smallest, the electric field strengths E_1, E_2, and E_3 at points 1, 2, and 3.

    Order:

    Explanation:

27 Current and Resistance

27.1 The Electron Current

1. A lightbulb is connected with wires to a battery, and the bulb is glowing. Are simple observations and measurements you can make on this circuit able to distinguish a current composed of positive charge carriers from a current composed of negative charge carriers? If so, describe how you can tell which it is. If not, why not?

2. One model of current consists of the motion of discrete, charged particles. Another model is that current is the flow of a continuous charged fluid. Do simple experiments provide evidence in favor of either one of these models? Explain.

3. Are the charge carriers always electrons? If so, why is this the case? If not, describe a situation in which a current is due to some other charge carrier.

27.2 Creating a Current

4. The electron drift speed in a wire is exceedingly slow—typically only a fraction of a millimeter per second. Yet when you turn on a light switch, a light bulb several meters away seems to come on instantly. Explain how to resolve this apparent paradox.

5. The figure shows a segment of a current-carrying metal wire.

   a. Is there an electric field inside the wire? If so, draw and label an arrow on the figure to show its direction. If not, why not?

   b. If there is an electric field, draw on the figure a possible arrangement of charges that could be the source charges causing the field.

6. a. If the electrons in a current-carrying wire collide with the positive ions *more* frequently, does their drift speed increase or decrease? Explain.

   b. Does an increase in the collision frequency make the wire a better conductor or a worse conductor? Explain.

   c. Would you expect a metal to be a better conductor at high temperature or at low temperature? Explain.

27.3 Current and Current Density

7. What is the difference between current and current density?

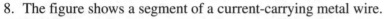

8. The figure shows a segment of a current-carrying metal wire.

   a. Draw an arrow on the figure, using a **black** pen or pencil, to show the direction of motion of the charge carriers.

   b. Draw an arrow on the figure, using a **red** pen or pencil, to show the direction of the electric field.

9. Is I_2 greater than, less than, or equal to I_1? Explain.

10. All wires in this figure are made of the same material and have the same diameter. Rank in order, from largest to smallest, the currents I_1 to I_4.

   Order:

   Explanation:

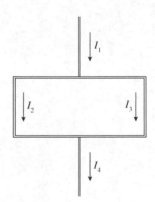

11. A lightbulb is connected to a battery with 1-mm-diameter wires. The bulb is glowing.

    a. Draw arrows at points 1, 2, and 3 to show the direction of the electric field at those points. (The points are *inside* the wire.)

    b. Rank in order, from largest to smallest, the field strengths E_1, E_2, and E_3.

    Order:

    Explanation:

12. A wire carries a 4 A current. What is the current in a second wire that delivers twice as much charge in half the time?

13. The current density in a wire is 1000 A/m². What will the current density be if the current is doubled and the wire's diameter is halved?

27.4 Conductivity and Resistivity

14. Metal 1 and metal 2 are each formed into 1-mm-diameter wires. The electric field needed to cause a 1 A current in metal 1 is larger than the electric field needed to cause a 1 A current in metal 2. Which metal has the larger conductivity? Explain.

15. If a metal is heated, does its conductivity increase, decrease, or stay the same? Explain.

16. Wire 1 and wire 2 are made from the same metal. Wire 1 has twice the diameter and half the electric field of wire 2. What is the ratio I_1/I_2?

17. Wire 1 and wire 2 are made from the same metal. Wire 2 has a larger diameter than wire 1. The electric field strengths E_1 and E_2 in the wires are equal.

    a. Compare the values of the two current densities. Is J_1 greater than, less than, or equal to J_2? Explain.

    b. Compare the values of the currents I_1 and I_2.

    c. Compare the values of the electron drift speeds $(v_d)_1$ and $(v_d)_2$.

18. A wire consists of two segments of different diameters but made from the same metal. The current in segment 1 is I_1.

    a. Compare the values of the currents in the two segments. Is I_2 greater than, less than, or equal to I_1? Explain.

    b. Compare the values of the current densities J_1 and J_2.

    c. Compare the strengths of the electric fields E_1 and E_2 in the two segments.

    d. Compare the values of the electron drift speeds $(v_d)_1$ and $(v_d)_2$.

19. A wire consists of two equal-diameter segments. Their conductivities differ, with $\sigma_2 > \sigma_1$. The current in segment 1 is I_1.

    a. Compare the values of the currents in the two segments. Is I_2 greater than, less than, or equal to I_1? Explain.

    b. Compare the strengths of the current densities J_1 and J_2.

    c. Compare the strengths of the electric fields E_1 and E_2 in the two segments.

27.5 Resistance and Ohm's Law

20. A continuous metal wire connects the two ends of a 3 V battery with a
    rectangular loop. The negative terminal of the battery has been chosen as the
    point where $V = 0$ V.

    a. Locate and label the approximate points along the wire where
       $V = 3$ V, $V = 2$ V, and $V = 1$ V.

    b. Points a and c are *inside* the wire. Point b is inside the battery. Does the
       electric field at a, b, and c point left, right, up, or down? Or is $\vec{E} = 0$?

    $\vec{E}_a$ _____ $\vec{E}_b$ _____ $\vec{E}_c$ _____

    c. In moving through the *wire* from the negative to the positive battery terminal, does the electric
       potential increase, decrease, or not change? If the potential changes, by how much does it change?

    d. In moving through the *battery* from the negative to the positive battery terminal, does the potential
       increase, decrease, or not change? If the potential changes, by how much does it change?

    e. In moving all the way around the loop in a clockwise direction, is the net change in the potential
       positive, negative, or zero?

21. a. Which direction—clockwise or counterclockwise—does an electron travel
       through the wire? Explain.

    b. Does an electron's electric potential energy increase, decrease, or stay the
       same as it moves through the wire? Explain.

    c. If you answered "decrease" in part b, where does the energy go? If you answered "increase" in
       part b, where does the energy come from?

22. The wires below are all made of the same material. Rank in order, from largest to smallest, the resistances R_1 to R_5 of these wires.

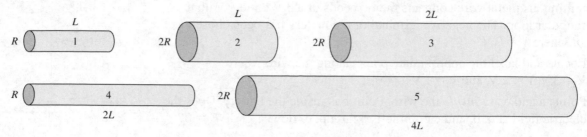

Order:

Explanation:

23. The two circuits use identical batteries and wires of equal diameters. Rank in order, from largest to smallest, the currents I_1, I_2, I_3, and I_4 at points 1 to 4.

Order:

Explanation:

24. The two circuits use identical batteries and wires of equal diameters. Rank in order, from largest to smallest, the currents I_1 to I_7 at points 1 to 7.

Order:

Explanation:

25. A wire is connected to the terminals of a 6 V battery. What is the potential difference ΔV_{wire} between the two ends of the wire, and what is the current I through the wire, if the wire has the following resistances:

6 V

a. $R = 1\ \Omega$ $\Delta V_{wire} =$ _____ $I =$ _____

b. $R = 2\ \Omega$ $\Delta V_{wire} =$ _____ $I =$ _____

c. $R = 3\ \Omega$ $\Delta V_{wire} =$ _____ $I =$ _____

d. $R = 6\ \Omega$ $\Delta V_{wire} =$ _____ $I =$ _____

26. The graph shows the current-versus-potential-difference relationship for a resistor R.

a. What is the numerical value of R?

b. Suppose the length of the resistor is doubled. On the figure, draw the current-versus-potential-difference graph for the longer resistor.

27. For resistors R_1 to R_2:

a. Which end (left, right, top, or bottom) is more positive?

R_1: _____ R_2: _____

b. In which direction (such as left to right or top to bottom) does the potential decrease?

R_1: _____

R_2: _____

28. Rank in order, from largest to smallest, the currents I_1 to I_4 through these four resistors.

Order:

Explanation:

29. Which, if any, of these statements are true? (More than one may be true.)

    i.   A battery supplies the energy to a circuit.
    ii. A battery is a source of potential difference. The potential difference between the terminals of the battery is always the same.
    iii. A battery is a source of current. The current leaving the battery is always the same.

    Explain your choice or choices.

28 Fundamentals of Circuits

28.1 Circuit Elements and Diagrams

28.2 Kirchhoff's Laws and the Basic Circuit

1. The tip of a flashlight bulb is touching the top of a 3 V battery. Does the bulb light? Why or why not?

2. Current I_{in} flows into three resistors connected together one after the other. The graph shows the value of the potential as a function of distance.

   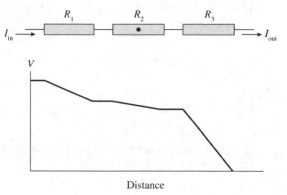

   a. Is I_{out} greater than, less than, or equal to I_{in}? Explain.

   b. Rank in order, from largest to smallest, the three resistances R_1, R_2, and R_3.

   Order:

   Explanation:

   c. Is there an electric field at the point inside R_2 that is marked with a dot? If so, in which direction does it point? If not, why not?

3. A flashlight bulb is connected to a battery and is glowing. Is current I_2 greater than, less than, or equal to current I_1? Explain.

4. a. In which direction does current flow through resistor R?

   b. Which end of R is more positive? Explain.

   c. If this circuit were analyzed in a clockwise direction, what numerical value would you assign to ΔV_{res}? Why?

   d. What value would ΔV_{res} have if the circuit were analyzed in a counterclockwise direction?

5. The wire is broken on the right side of this circuit. What is the potential difference ΔV_{12} between points 1 and 2? Explain.

6. Draw a circuit for which the Kirchhoff's loop law equation is

$$6V - I \cdot 2\Omega + 3V - I \cdot 4\Omega = 0$$

Assume that the analysis is done in a clockwise direction.

7. The current in a circuit is 2.0 A. The graph shows how the potential changes when going around the circuit in a clockwise direction, starting from the lower left corner. Draw the circuit diagram.

28.3 Energy and Power

8. This circuit has two resistors, with $R_1 > R_2$. Which of the two resistors dissipates the larger amount of power? Explain.

9. A 60 W lightbulb and a 100 W lightbulb are placed one after the other in a circuit. The battery's emf is large enough that both bulbs are glowing. Which one glows more brightly? Explain.

10. Two conductors of equal lengths are connected to a battery by ideal wires. The conductors are made of the same material but have different radii r. Which of the two conductors dissipates the larger amount of power? Explain.

11. The circuit shown has a battery and two resistors, with $R_1 > R_2$. Which of the two resistors dissipates the larger amount of power? Explain your reasoning.

28.4 Series Resistors

28.5 Real Batteries

12. What is the equivalent resistance of each group of resistors?

a.

2 Ω

3 Ω

6 Ω

$R_{eq} =$ _____

b.

3 Ω

3 Ω

3 Ω

$R_{eq} =$ _____

c.

2 Ω   1 Ω   2 Ω

$R_{eq} =$ _____

13. The figure shows two circuits. The two ideal batteries are identical and the four resistors all have exactly the same resistance.

a. Is ΔV_{ab} larger than, smaller than, or equal to ΔV_{cd}? Explain.

b. Rank in order, from largest to smallest, the currents I_1, I_2, and I_3.

Order:

Explanation:

14. The lightbulb in this circuit has a resistance of 1 Ω.

a. What are the values of:

2 Ω

3 V

ΔV_{12} _____

ΔV_{23} _____

ΔV_{34} _____

b. Suppose the bulb is now removed from its socket. Then what are the values of:

ΔV_{12} _____

ΔV_{23} _____

ΔV_{34} _____

15. If the value of R is increased, does ΔV_{bat} increase, decrease, or stay the same? Explain.

28.6 Parallel Resistors

16. What is the equivalent resistance of each group of resistors?

a.

$R_{eq} =$ _____

b.

$R_{eq} =$ _____

c.

2 Ω
1 Ω
2 Ω

$R_{eq} =$ _____

17. a. What fraction of current I goes through the 3 Ω resistor?

3 Ω

$I \rightarrow$

9 Ω

b. If the 9 Ω resistor is replaced with a larger resistor, will the fraction of current going through the 3 Ω resistor increase, decrease, or stay the same?

18. The figure shows five combinations of identical resistors. Rank in order, from largest to smallest, the equivalent resistances $(R_{eq})_1$ to $(R_{eq})_5$.

1   2   3   4   5

Order:

Explanation:

28.7 Resistor Circuits

28.8 Getting Grounded

19. Rank in order, from largest to smallest, the three currents I_1 to I_3.

    Order:

    Explanation:

20. The two ideal batteries are identical and the four resistors all have exactly the same resistance.

    a. Compare ΔV_{ab}, ΔV_{cd}, and ΔV_{ef}. Are they all the same? If not, rank them in decreasing order. Explain your reasoning.

    b. Rank in order, from largest to smallest, the five currents I_1 to I_5.

    Order:

    Explanation:

21. Bulbs A and B are identical and initially both are glowing. Then the switch is closed. What happens to each bulb? Does its brightness increase, stay the same, decrease, or go out? Explain.

Exercises 22–25: Assume that all wires are ideal (zero resistance) and that all batteries are ideal (constant potential difference).

22. Initially bulbs A and B are glowing. Then the switch is closed. What happens to each bulb? Does it get brighter, stay the same, get dimmer, or go out? Explain your reasoning.

23. a. Bulbs A, B, and C are identical. Rank in order, from most to least, the brightnesses of the three bulbs.

Order:

Explanation:

b. Suppose a wire is connected between points 1 and 2. What happens to each bulb? Does it get brighter, stay the same, get dimmer, or go out? Explain.

24. a. Consider the points a and b. Is the potential difference $\Delta V_{ab} = 0$? If so, why? If not, which point is more positive?

b. If a wire is connected between points a and b, does a current flow through it? If so, in which direction—to the right or to the left? Explain.

25. Bulbs A and B are identical. Initially both are glowing.

a. Bulb A is removed from its socket. What happens to bulb B? Does it get brighter, stay the same, get dimmer, or go out? Explain.

b. Bulb A is replaced. Bulb B is then removed from its socket. What happens to bulb A? Does it get brighter, stay the same, get dimmer, or go out? Explain.

c. The circuit is restored to its initial condition. A wire is then connected between points 1 and 2. What happens to the brightness of each bulb?

26. Real circuits frequently have a need to reduce a voltage to a smaller
PSS value. This is done with a *voltage-divider circuit* such as the one
28.1 shown here. We've shown the input voltage V_{in} as a battery, but in
practice, it might be a voltage signal produced by some other circuit,
such as the receiver circuit in your television.

a. Is this a series or a parallel circuit? _____

b. Symbolically, what is the value of the equivalent resistance? $R_{eq} = $ _____

c. Redraw the circuit with the two resistors replaced with the equivalent resistance.

d. You now have one resistor connected directly across the battery. Find the voltage across and the
current through this resistor. Write your answers in terms of V_{in}, R_1, and R_2.

$\Delta V = $ _____ $I = $ _____

e. Which of these, ΔV or I, is the same for R_1 and R_2 as for R_{eq}? _____

f. Using Ohm's law and your answer to e, what is V_{out}? It's most useful to write your answer as
$V_{out} = $ something $\times V_{in}$.

$V_{out} = $ _____

g. What is V_{out} if $R_2 = 2R_1$? $V_{out} = $ _____

28.9 *RC* Circuits

27. The graph shows the voltage as a function of time on a capacitor as it is discharged (separately) through three different resistors. Rank in order, from largest to smallest, the values of the resistances R_1 to R_3.

Order:

Explanation:

28. The capacitors in each circuit are fully charged before the switch closes. Rank in order, from largest to smallest, the time constants τ_1 to τ_5 with which the capacitors will discharge after the switch closes.

Order:

Explanation:

29. The charge on the capacitor is zero when the switch closes at
    $t = 0$ s.

    a. What will be the current in the circuit after the switch has
       been closed for a long time? Explain.

    b. Immediately after the switch closes, before the capacitor has had time to charge, the potential
       difference across the capacitor is zero. What must be the potential difference across the resistor
       in order to satisfy Kirchhoff's loop law? Explain.

    c. Based on your answer to part b, what is the current in the circuit immediately after the switch closes?

    d. Sketch a graph of current versus time, starting from just
       before $t = 0$ s and continuing until the switch has been
       closed a long time. There are no numerical values for
       the horizontal axis, so you should think about the *shape*
       of the graph.

29 The Magnetic Field

29.1 Magnetism

1. A lightweight glass sphere hangs by a thread. The north pole of a bar magnet is brought near the sphere.

   a. Suppose the sphere is electrically neutral. How does it respond?

      i. It is strongly attracted to the magnet.

      ii. It is weakly attracted to the magnet.

      iii. It does not respond.

      iv. It is weakly repelled by the magnet.

      v. It is strongly repelled by the magnet.

      Explain your choice.

   b. How does the sphere respond if it is positively charged? Explain.

2. A metal sphere hangs by a thread. When the north pole of a bar magnet is brought near, the sphere is strongly attracted to the magnet. Then the magnet is reversed and its south pole is brought near the sphere. How does the sphere respond? Explain.

3. The compass needle below is free to rotate in the plane of the page. Either a bar magnet or a charged rod is brought toward the *center* of the compass. Does the compass rotate? If so, in which direction? If not, why not?

a.

b.

4. You have two electrically neutral metal cylinders that exert strong attractive forces on each other. You have no other metal objects. Can you determine if *both* of the cylinders are magnets, or if one is a magnet and the other just a piece of iron? If so, how? If not, why not?

5. Can you think of any kind of object that is repelled by *both* ends of a bar magnet? If so, what? If not, what prevents this from happening?

29.2 The Discovery of the Magnetic Field

6. A neutral copper rod, a polarized insulator, and a bar magnet are arranged around a current-carrying wire as shown. For each, will it stay where it is? Move toward or away from the wire? Rotate clockwise or counterclockwise? Explain.

    a. Neutral copper rod

    b. Polarized insulator

    c. Bar magnet

7. For each of the current-carrying wires shown, draw a compass needle in its equilibrium position at the positions of the dots. Label the poles of the compass needle.

    a.   b.

8. The figure shows a wire directed into the page and a nearby compass needle. Is the wire's current going into the page or coming out of the page? Explain.

9. A compass is placed at 12 different positions and its orientation is recorded. Use this information to draw the magnetic *field lines* in this region of space. Draw the field lines on the figure.

29.3 The Source of the Magnetic Field: Moving Charges

10. A positively charged particle moves toward the bottom of the page.

    a. At each of the six number points, show the direction of the magnetic field or, if appropriate, write $\vec{B} = 0$.

    b. Rank in order, from strongest to weakest, the magnetic field strengths B_1 to B_6 at these points.

    Order:

    Explanation:

11. The negative charge is moving out of the page, coming toward you. Draw the magnetic field lines in the plane of the page.

12. Two charges are moving as shown. At this instant of time, the net magnetic field at point 2 is $\vec{B}_2 = 0$.

    a. Is the unlabeled moving charge positive or negative? Explain.

    b. What is the magnetic field direction at point 1? Explain.

    c. What is the magnetic field direction at point 3?

29.4 The Magnetic Field of a Current

29.5 Magnetic Dipoles

13. Each figure shows a current-carrying wire. Draw the magnetic field diagram:

a.

b.

The wire is perpendicular to the page. Draw magnetic field *lines*, then show the magnetic field *vectors* at a few points around the wire.

The wire is in the plane of the page. Show the magnetic field above and below the wire.

14. This current-carrying wire is in the plane of the page. Draw the magnetic field on both sides of the wire.

15. Use an arrow to show the current direction in this wire.

16. Each figure below shows two long straight wires carrying equal currents into or out of the page. At each of the dots, use a **black** pen or pencil to show and label the magnetic fields $\vec{B}_1$ and $\vec{B}_2$ due to each wire. Then use a **red** pen or pencil to show the net magnetic field.

a.

Wire 1
⊗

•   •

⊙
Wire 2

b.

Wire 1
⊙

•   •

⊙
Wire 2

17. A long straight wire, perpendicular to the page, passes through a uniform magnetic field. The *net* magnetic field at point 3 is zero.

    a. On the figure, show the direction of the current in the wire.

    b. Points 1 and 2 are the same distance from the wire as point 3, and point 4 is twice as distant. Construct vector diagrams at points 1, 2, and 4 to determine the net magnetic field at each point.

18. A long straight wire passes above one edge of a current loop. Both are perpendicular to the page. $\vec{B}_1 = \vec{0}$ at point 1.

    a. On the figure, show the direction of the current in the loop.

    b. Use a vector diagram to determine the net magnetic field at point 2.

19. The figure shows the magnetic field seen when facing a current loop in the plane of the page.

    a. On the figure, show the direction of the current in the loop.

    b. Is the north pole of this loop at the upper surface of the page or the lower surface of the page? Explain.

20. The current loop exerts a repulsive force on the bar magnet. On the figure, show the direction of the current in the loop. Explain.

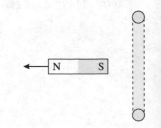

29.6 Ampère's Law and Solenoids

21. What is the total current through the area bounded by the closed curve?

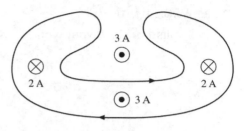

22. The total current through the area bounded by the closed curve is 2 A. What are the size and direction of I_3?

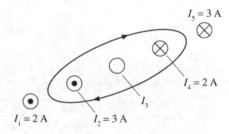

23. The magnetic field above the dotted line is $\vec{B} = (2\ \text{T, right})$. Below the dotted line the field is $\vec{B} = (2\ \text{T, left})$. Each closed loop is $1\ \text{m} \times 1\ \text{m}$. Let's evaluate the line integral of $\vec{B}$ around each of these closed loops by breaking the integration into four steps. We'll go around the loop in a *clockwise* direction. Pay careful attention to signs.

	Loop 1	Loop 2	Loop 3
$\int \vec{B} \cdot d\vec{s}$ along left edge			
$\int \vec{B} \cdot d\vec{s}$ along top			
$\int \vec{B} \cdot d\vec{s}$ along right edge			
$\int \vec{B} \cdot d\vec{s}$ along bottom			

The line integral *around* the loop is simply the sum of these four separate integrals:

$\oint \vec{B} \cdot d\vec{s}$ around the loop

24. The strength of a circular magnetic field decreases with increasing radius as shown.

    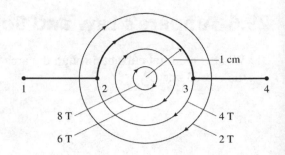

    a. What is $\int_1^2 \vec{B} \cdot d\vec{s}$? _____

    Explain or show your work.

    b. What is $\int_2^3 \vec{B} \cdot d\vec{s}$? _____   Explain or show your work.

    c. What is $\int_3^4 \vec{B} \cdot d\vec{s}$? _____   Explain or show your work.

    d. Combining your answers to parts a to c, what is $\int_1^4 \vec{B} \cdot d\vec{s}$? _____

25. A solenoid with one layer of turns produces the magnetic field strength you need for an experiment when the current in the coil is 3 A. Unfortunately, this amount of current overheats the coil. You've determined that a current of 1 A would be more appropriate. How many additional layers of turns must you add to the solenoid to maintain the magnetic field strength? Explain.

26. Rank in order, from largest to smallest, the magnetic fields B_1 to B_3 produced by these three solenoids.

    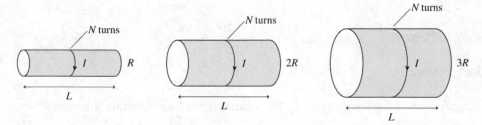

    Order:

    Explanation:

29.7 The Magnetic Force on a Moving Charge

27. For each of the following, draw the magnetic force vector on the charge or, if appropriate, write "$\vec{F}$ into page," "$\vec{F}$ out of page," or "$\vec{F} = \vec{0}$."

a.

b.

c.

d.

$\vec{v}$ out of page

e.

f.

28. For each of the following, determine the sign of the charge (+ or −).

a.

$\vec{F}$ into page

b.

$\vec{v}$ into page

c.

d.

 a. $q =$ _____ b. $q =$ _____ c. $q =$ _____ d. $q =$ _____

29. The magnetic field has constant magnitude inside the dashed lines and is zero outside. Sketch and label the trajectory of the charge for

 a. A weak field.

 b. A strong field.

30. A positive ion, initially traveling into the page, is shot through the gap in a horseshoe magnet. Is the ion deflected up, down, left, or right? Explain.

31. A positive ion is shot between the plates of a parallel-plate capacitor.

    a. In what direction is the electric force on the ion?

    b. Could a magnetic field exert a magnetic force on the ion that is opposite in direction to the electric force? If so, show the magnetic field on the figure.

32. In a high-energy physics experiment, a neutral particle enters a bubble chamber in which a magnetic field points into the page. The neutral particle undergoes a collision inside the bubble chamber, creating two charged particles. The subsequent trajectories of the charged particles are shown.

    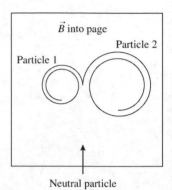

    a. What is the sign (+ or −) of particle 1? _____

    What is the sign (+ or −) of particle 2? _____

    b. Which charged particle leaves the collision with a larger momentum? Explain. (Assume that $|q| = e$ for both particles.)

33. A hollow solenoid is wound as shown and attached to a battery. Two electrons are fired into the solenoid, one from the end and one through a very small hole in the side.

    a. In what direction does the magnetic field inside the solenoid point? Show it on the figure.

    b. Is electron 1 deflected as it moves through the solenoid? If so, in which direction? If not, why not?

    c. Is electron 2 deflected as it moves through the solenoid? If so, in which direction? If not, why not?

34. Two protons are traveling in the directions shown.

    a. Draw and label the electric force on each proton due to the other proton.

    b. Draw and label the magnetic force on each proton due to the other proton. Explain how you determined the directions.

29.8 Magnetic Forces on Current-Carrying Wires

29.9 Forces and Torques on Current Loops

35. Three current-carrying wires are perpendicular to the page. Construct a force vector diagram on the figure to find the net force on the upper wire due to the two lower wires.

36. Three current-carrying wires are perpendicular to the page.

   a. Construct a force vector diagram on each wire to determine the direction of the net force on each wire.

   b. Can three *charges* be placed in a triangular pattern so that their force diagram looks like this? If so, draw it below. If not, why not?

37. A current-carrying wire passes in front of a solenoid that is wound as shown. The wire experiences an upward force. Use arrows to show the direction in which the current enters and leaves the solenoid. Explain your choice.

38. A current loop is placed between two bar magnets. Does the loop move to the right, move to the left, rotate clockwise, rotate counterclockwise, some combination of these, or none of these? Explain.

39. A square current loop is placed in a magnetic field as shown.

    a. Does the loop undergo a displacement? If so, is it up, down, left, or right? If not, why not?

    b. Does the loop rotate? If so, which edge rotates out of the page and which edge into the page? If not, why not?

40. The south pole of a bar magnet is brought toward the current loop. Does the bar magnet attract the loop, repel the loop, or have no effect on the loop? Explain.

29.10 Magnetic Properties of Matter

41. A solenoid, wound as shown, is placed next to an unmagnetized piece of iron. Then the switch is closed.

    a. Identify on the figure the north and south poles of the solenoid.

    b. What is the direction of the solenoid's magnetic field as it passes through the iron?

    c. What is the direction of the induced magnetic dipole in the iron?

    d. Identify on the figure the north and south poles of the induced magnetic dipole in the iron.

    e. When the switch is closed, does the iron move left or right? Does it rotate? Explain.

30 Electromagnetic Induction

30.1 Induced Currents

30.2 Motional emf

1. The figures below show one or more metal wires sliding on fixed metal rails in a magnetic field. For each, determine if the induced current flows clockwise, flows counterclockwise, or is zero. Show your answer by drawing it.

a.

b.

c.

d.

e.

f.

2. A loop of copper wire is being pulled from between two magnetic poles.

   a. Show on the figure the current induced in the loop. Explain your reasoning.

   b. Does either side of the loop experience a magnetic force? If so, draw and label a vector arrow or arrows on the figure to show any forces.

3. A vertical, rectangular loop of copper wire is half in and half out of a horizontal magnetic field. (The field is zero beneath the dashed line.) The loop is released and starts to fall.

   a. Add arrows to the figure to show the direction of the induced current in the loop.

   b. Is there a net magnetic force on the loop? If so, in which direction? Explain.

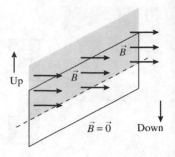

4. Two very thin sheets of copper are pulled through a magnetic field. Do eddy currents flow in the sheet? If so, show them on the figures, with arrows to indicate the direction of flow. If not, why not?

   a.

   b.

5. The figure shows an edge view of a copper sheet being pulled between two magnetic poles.

   a. Add a dot or an ✕ to each of the circles to indicate the direction in which eddy currents are flowing in and out of the page.

   b. Do the currents you labeled in part a experience magnetic forces? If so, add force vectors to the figure to show the directions. If not, why not?

   c. Is there a net magnetic force on the copper sheet? If so, in which direction?

30.3 Magnetic Flux

6. The figure shows five loops in a magnetic field. The numbers indicate the lengths of the sides and the strength of the field. Rank in order, from largest to smallest, the magnetic fluxes Φ_1 to Φ_5. Some may be equal.

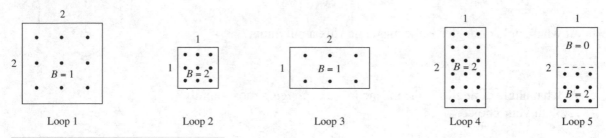

Loop 1   Loop 2   Loop 3   Loop 4   Loop 5

Order:

Explanation:

7. The figure shows four circular loops that are perpendicular to the page. The radius of loops 3 and 4 is twice that of loops 1 and 2. The magnetic field is the same for each. Rank in order, from largest to smallest, the magnetic fluxes Φ_1 to Φ_4. Some may be equal.

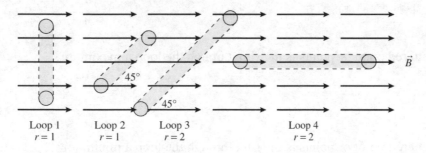

Loop 1   Loop 2   Loop 3   Loop 4
$r = 1$ $r = 1$ $r = 2$ $r = 2$

Order:

Explanation:

8. A circular loop rotates at constant speed about an axle through the center of the loop. The figure shows an edge view and defines the angle ϕ, which increases from 0° to 360° as the loop rotates.

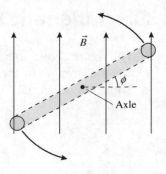

   a. At what angle or angles is the magnetic flux a maximum?

   b. At what angle or angles is the magnetic flux a minimum?

   c. At what angle or angles is the magnetic flux *changing* most rapidly? Explain your choice.

9. A magnetic field is perpendicular to a loop. The graph shows how the magnetic field changes as a function of time, with positive values for B indicating a field into the page and negative values a field out of the page. Several points on the graph are labeled.

Field through loop

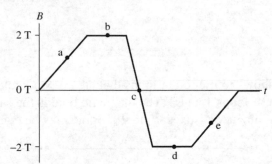

   a. At which lettered point or points is the flux through the loop a maximum?

   b. At which lettered point or points is the flux through the loop a minimum?

   c. At which point or points is the flux changing most rapidly?

   d. At which point or points is the flux changing least rapidly?

30.4 Lenz's Law

30.5 Faraday's Law

10. Does the loop of wire have a clockwise current, a counterclockwise current, or no current under the following circumstances? Explain.

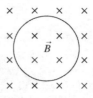

a. The magnetic field points into the page and its strength is increasing.

b. The magnetic field points into the page and its strength is constant.

c. The magnetic field points into the page and its strength is decreasing.

11. A loop of wire is perpendicular to a magnetic field. The magnetic field strength as a function of time is given by the top graph. Draw a graph of the current in the loop as a function of time. Let a positive current represent a current that comes out of the top and enters the bottom. There are no numbers for the vertical axis, but your graph should have the correct shape and proportions.

12. A loop of wire is horizontal. A bar magnet is pushed toward the loop from below, along the axis of the loop.

    a. What is the current direction in the loop as the magnet is approaching? Explain.

    b. Is there a magnetic force on the loop? If so, in which direction? Explain.
       Hint: A current loop is a magnetic dipole.

13. A bar magnet is dropped, south pole down, through the center of a loop of wire. The center of the magnet passes the plane of the loop at time t_c.

    a. Sketch a graph of the magnetic flux through the loop as a function of time.

    b. Sketch a graph of the current in the loop as a function of time. Let a clockwise current be a positive number and a counterclockwise current be a negative number.

14. a. Just after the switch on the left coil is closed, does current flow right to left or left to right through the current meter of the right coil? Or is the current zero? Explain.

    b. Long after the switch on the left coil is closed, does current flow right to left or left to right through the current meter of the right coil? Or is the current zero? Explain.

15. A solenoid is perpendicular to the page, and its field strength is increasing. Three circular wire loops of equal radii are shown. Rank in order, from largest to smallest, the size of the induced emf in the three rings.

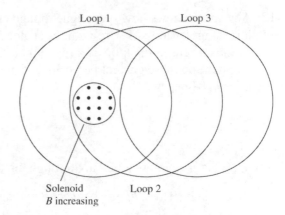

Solenoid
B increasing

> Order:
>
> Explanation:

16. A conducting loop around a magnetic field contains two lightbulbs, A and B. The wires connecting the bulbs are ideal, with no resistance. The magnetic field is increasing rapidly.

    a. Do the bulbs glow? Why or why not?

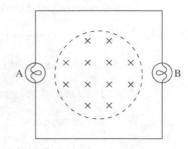

    b. If they glow, which bulb is brighter? Or are they equally bright? Explain.

17. A metal wire is resting on a U-shaped conducting rail. The rail is fixed in position, but the wire is free to move.

    a. If the magnetic field is increasing in strength, does the wire:

Fixed rail   Wire

    i. Remain in place?
    ii. Move to the right?
    iii. Move to the left?
    iv. Move up on the page?
    v. Move down on the page?

    vi. Move out of the plane of the page, breaking contact with the rail?
    vii. Rotate clockwise?
    viii. Rotate clockwise?
    ix. Some combination of these? If so, which?

    Explain your choice.

    b. If the magnetic field is decreasing in strength, which of the above happens? Explain.

18. The graph shows how the magnetic field changes through
PSS a rectangular loop of wire with resistance R. Draw a graph
30.1 of the current in the loop as a function of time. Let a
counterclockwise current be positive, a clockwise
current be negative.

a. What is the magnetic flux through the loop at $t = 0$?

b. Does this flux *change* between $t = 0$ and $t = t_1$?

c. Is there an induced current in the loop between $t = 0$ and $t = t_1$?

d. What is the magnetic flux through the loop at $t = t_2$?

e. What is the *change* in flux through the loop between t_1 and t_2?

f. What is the time interval between t_1 and t_2?

g. What is the magnitude of the induced emf between t_1 and t_2?

h. What is the magnitude of the induced current between t_1 and t_2?

i. Does the magnetic field point out of or into the loop?

j. Between t_1 and t_2, is the magnetic flux increasing or decreasing?

k. To oppose the *change* in the flux between t_1 and t_2, should the magnetic
   field of the induced current point out of or into the loop?

l. Is the induced current between t_1 and t_2 positive or negative?

m. Does the flux through the loop change after t_2?

n. Is there an induced current in the loop after t_2?

o. Use all this information to draw a graph of the induced current. Add appropriate labels on the
   vertical axis.

30.6 Induced Fields

19. Consider these two situations:

    a. Draw the induced electric field.

    $\vec{B}$-field rapidly increasing

    b. Draw the induced electric field.

    $\vec{B}$-field rapidly decreasing

30.7 Induced Current: Three Applications

No exercises.

30.8 Inductors

20. The figure shows the current through an inductor. Draw a graph showing the potential difference ΔV_L across the inductor. There are no numbers, but your graph should have the correct shape and proportions.

21. The figure shows the potential difference across an inductor. There is no current at $t = 0$. Draw a graph of the current through the inductor as a function of time. There are no numbers, but your graph should have the correct shape and proportions.

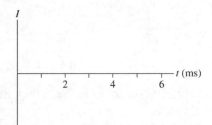

30.9 *LC* Circuits

22. An *LC* circuit oscillates at a frequency of 2000 Hz. What will the frequency be if the inductance is quadrupled?

23. The capacitor in an *LC* circuit has maximum charge at $t = 1$ μs. The current through the inductor next reaches a maximum at $t = 3$ μs.

    a. When will the inductor current reach a maximum in the opposite direction?

    b. What is the circuit's period of oscillation?

30.10 *LR* Circuits

24. Three *LR* circuits are made with the same resistor but different inductors. The figure shows the inductor current as a function of time. Rank in order, from largest to smallest, the three inductances L_1, L_2, and L_3.

    Order:
    Explanation:

25. a. What is the battery current immediately after the switch closes? Explain.

    b. What is the battery current after the switch has been closed a long time? Explain.

31 Electromagnetic Fields and Waves

31.1 *E* or *B*? It Depends on Your Perspective

1. In frame A, a positive charge moves through the magnetic field shown.

   a. Draw a vector on the charge to show the magnetic force in A.

   b. What are the speed v_{BA} and direction of a reference frame B in which there is no magnetic force? Explain.

   c. What are the type and direction of any fields in B that could cause the observed force on the charge?

2. Sharon drives her rocket through a magnetic field, traveling to the right at a speed of 1000 m/s as measured by Bill. As she passes Bill, she shoots a positive charge backward at a speed of 1000 m/s relative to her.

   a. According to Bill, what kind of force or forces act on the charge? In which directions? Explain.

   b. According to Sharon, what kind of force or forces act on the charge? In which directions? Draw the forces on the charge.

3. In frame A, a positive charge Q moves to the right with velocity v_{OA}. Frame B travels to the right at $v_{BA} = v_{OA}$ relative to A. Frame C travels to the right at $v_{CA} = 2v_{OA}$ relative to A. The figure below shows the charge three times, once in each reference frame.

   a. For each:
      • Draw and label a velocity vector on the charge showing its motion in that frame.
      • Draw and label the electric and magnetic field vectors <u>due to the charge</u> at the points marked with small dots above and below the charge. Use the notation of circled × and • to show fields into or out of the page.

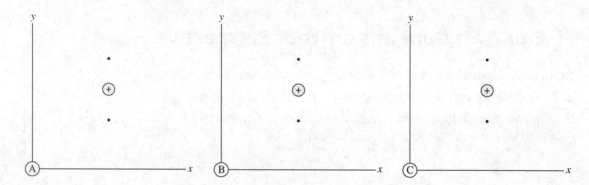

   b. Does it make sense to talk about "the" magnetic field? Why or why not?

31.2 The Field Laws Thus Far

31.3 The Displacement Current

4. If you curl the fingers of your right hand as shown, is the electric flux positive or negative?

a.

b.

Sign of Φ_e _____   Sign of Φ_e _____

5. If you curl the fingers of your right hand as shown, is the emf positive or negative?

Induced current

6. What is the current through surface S?

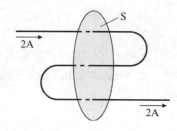

7. The capacitor in this circuit was initially charged, then the switch was closed. At this instant of time, the potential difference across the resistor is $\Delta V_R = 4\text{ V}$.

a. At this instant of time, what is the current through the resistor?

b. At this instant of time, what is the current through the space between the capacitor plates?

c. At this instant of time, what is the displacement current through the space between the capacitor plates?

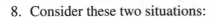

d. Is the displacement current really a current? If so, what are the moving charges? If not, what is the displacement current?

8. Consider these two situations:

a.

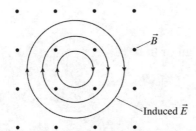

Is the magnetic field strength increasing, decreasing, or not changing? Explain.

b.

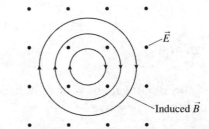

Is the electric field strength increasing, decreasing, or not changing? Explain.

9. Consider these two situations:

a. Draw the induced electric field.

×　×　×　×

×　×　×　×

×　×　×　×

×　×　×　×

$\vec{B}$-field rapidly increasing

b. Draw the induced magnetic field.

×　×　×　×

×　×　×　×

×　×　×　×

×　×　×　×

$\vec{E}$-field rapidly increasing

31.4 Maxwell's Equations

31.5 Electromagnetic Waves

31.6 Properties of Electromagnetic Waves

10. This is an electromagnetic plane wave traveling into the page.
    Draw the magnetic field vectors $\vec{B}$ at the dots.

11. This is an electromagnetic wave at one instant of time.
    a. Draw the velocity vector $\vec{v}_{em}$.

    b. Draw $\vec{E}$, $\vec{B}$, and $\vec{v}_{em}$ a half cycle later.

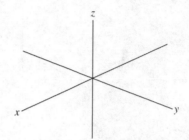

12. Do the following represent possible electromagnetic waves? If not, why not?

a.

b.

c.

d.

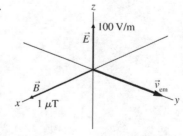

13. The intensity of an electromagnetic wave is 10 W/m^2. What will be the intensity if:

    a. The amplitude of the electric field is doubled?

    b. The frequency is doubled?

31.7 Polarization

14. A polarized electromagnetic wave passes through a polarizing filter. Draw the electric field of the wave after it has passed through the filter.

    a.   b.

15. A polarized electromagnetic wave passes through a series of polarizing filters. Draw the electric field of the wave after it has passed through each filter.

    a.

    b.

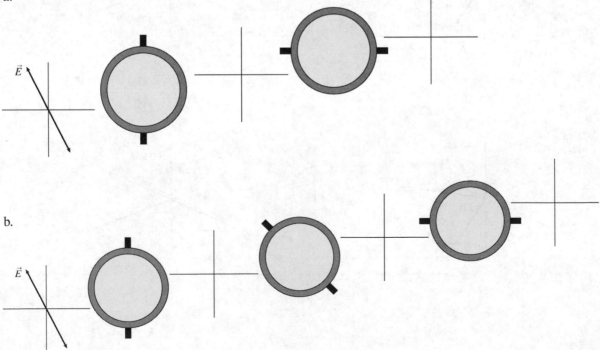

32 AC Circuits

32.1 AC Sources and Phasors

1. The figure shows emf phasors A, B, and C.

   a. What is the instantaneous value of the emf?

   A _____ B _____ C _____

   b. At this instant, is the magnitude of the emf increasing, decreasing, or holding constant?

   A _____ B _____ C _____

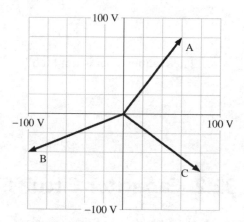

2. Draw a phasor diagram for the following emfs.

   a. $(100 \text{ V}) \cos \omega t$ at $\omega t = 240°$ b. $(400 \text{ V}) \cos \omega t$ at $t = \frac{1}{3}T$ c. $(200 \text{ V}) \cos \omega t$ at $t = 0$

3. The current phasor is shown for a 10 Ω resistor.

   a. What is the instantaneous resistor voltage v_R?

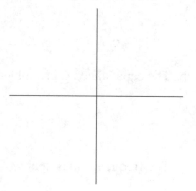

   b. What is the peak resistor voltage V_R?

4. A circuit consists of one resistor connected to an emf. The peak current through the resistor is 4.0 A. What is the peak current if:

   a. The resistance R is doubled?

   b. The peak emf $\mathcal{E}_0$ is doubled?

   c. The frequency ω is doubled?

32.2 Capacitor Circuits

5. A circuit consists of one capacitor connected to an emf. The peak current through the capacitor is 4.0 A. What is the peak current if:

   a. The peak emf $\mathcal{E}_0$ is doubled?

   b. The capacitance C is doubled?

   c. The frequency ω is doubled?

6. Current and voltage graphs are shown for a capacitor circuit with $\omega = 1000$ rad/s.

   a. What is the capacitive reactance X_C?

   b. What is the capacitance C?

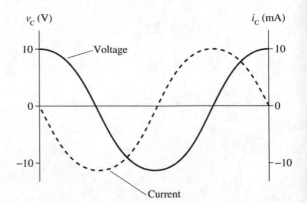

7. A 13 μF capacitor is connected to a 5.5 V/250 Hz oscillator. What is the instantaneous capacitor current i_C when $\mathcal{E}=-5.5$ V?

8. Consider these three circuits.

1   2   3

Rank in order, from largest to smallest, the peak currents $(I_C)_1$ to $(I_C)_3$ provided by the emf.

Order:

Explanation:

9. Consider these four circuits.

1   2   3   4

Rank in order, from largest to smallest, the capacitive reactances $(X_C)_1$ to $(X_C)_4$.

Order:

Explanation:

32.3 *RC* Filter Circuits

10. A low-pass *RC* filter has a crossover frequency $f_c = 200$ Hz. What is f_c if:

    a. The resistance R is halved?

    b. The capacitance C is halved?

    c. The peak emf $\mathcal{E}_0$ is halved?

11. What new resistor value R will give this circuit the same value of ω_c if the capacitor value is changed to:

    a. $C = 1\ \mu F$ $R =$ _____

    b. $C = 4\ \mu F$ $R =$ _____

    c. $C = 20\ \mu F$ $R =$ _____

12. Consider these three circuits.

    Rank in order, from largest to smallest, the crossover frequencies ω_{c1} to ω_{c3}.

    Order:

    Explanation:

13. The text claims that $V_R = V_C = \mathcal{E}_0/\sqrt{2}$ at $\omega = \omega_c$. If this is true, then $V_R + V_C > \mathcal{E}_0$. Is it possible for their sum to be larger than $\mathcal{E}_0$? Explain.

32.4 Inductor Circuits

14. A circuit consists of one inductor connected to an emf. The peak current through the inductor is 4.0 A. What is the peak current if:

a. The peak emf $\mathcal{E}_0$ is doubled?

b. The inductance L is doubled?

c. The frequency ω is doubled?

15. Current and voltage graphs are shown for an inductor circuit with $\omega = 1000$ rad/s.

a. What is the inductive reactance X_L?

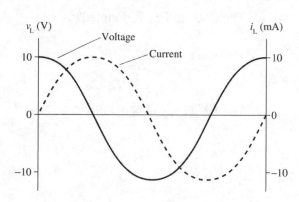

b. What is the inductance L?

16. Consider these four circuits.

Rank in order, from largest to smallest, the inductive reactances $(X_L)_1$ to $(X_L)_4$.

Order:

Explanation:

32.5 The Series *RLC* Circuit

17. The resonance frequency of a series *RLC* circuit is 1000 Hz. What is the resonance frequency if:
    a. The resistance *R* is doubled?

    b. The inductance *L* is doubled?

    c. The capacitance *C* is doubled?

    d. The peak emf $\mathcal{E}_0$ is doubled?

    e. The frequency ω is doubled?

18. For these combinations of resistance and reactance, is a series *RLC* circuit in resonance (Yes or No)? Does the current lead the emf, lag the emf, or is it in phase with the emf?

R	X_L	X_C	Resonance?	Current?
100 Ω	100 Ω	50 Ω		
100 Ω	50 Ω	100 Ω		
100 Ω	75 Ω	75 Ω		

19. In this series *RLC* circuit, is the emf frequency less than, equal to, or greater than the resonance frequency ω_0? Explain.

20. The emf frequency of a series *RLC* circuit is less than the circuit's resonance frequency. Does the current lead or lag the emf? Explain.

21. Consider these four circuits. They all have the same resonance frequency ω_0.

| 1 | 2 | 3 | 4 |

Rank in order, from largest to smallest, the maximum currents $(I_{max})_1$ to $(I_{max})_4$.

Order:

Explanation:

32.6 Power in AC Circuits

22. The current in a series *RLC* circuit lags the emf by 20°. You cannot change the emf. What two different things could you do to the circuit that would increase the power delivered to the circuit by the emf?

23. An average power dissipated by a resistor in an all-resistor circuit is 4.0 W. What is P_{avg} if:

    a. The resistance R is doubled?

    b. The peak emf $\mathcal{E}_0$ is doubled?

    c. The frequency is doubled?

24. Consider these two circuits.

    a. Which has the larger peak current? Or do both have the same peak current? Explain.

    b. Which emf supplies the larger power? Or do both supply the same power?

33 Wave Optics

33.1 Models of Light

33.2 The Interference of Light

1. The figure shows the light intensity recorded by a light-sensitive detector in an interference experiment. Notice that the light intensity comes "full on" at the edges of each maximum, so this is *not* the intensity that would be recorded in Young's double-slit experiment.

   a. Draw a graph of light intensity versus position on the detector. Your graph should have the same horizontal scale as the "photograph" above it.
   b. Is it possible to tell, from the information given, what the wavelength of the light is? If so, what is it? If not, why not?

2. The graph shows the light intensity on the viewing screen during a double-slit interference experiment.

   a. Draw the "photograph" that would be recorded if a light-sensitive detector were placed at the position of the screen. Your "photograph" should have the same horizontal scale as the graph above it. Be as accurate as you can. Let the white of the paper be the brightest intensity and a very heavy pencil shading be the darkest.
   b. Three positions on the screen are marked as A, B, and C. Draw history graphs showing the electric field amplitude of the light wave at each of these three positions as a function of time. Show three cycles, and use the same vertical scale on all three.

Photograph

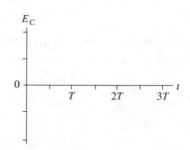

3. The figure below is a double-slit experiment seen looking down on the experiment from above. Although we usually see the light intensity only on a viewing screen, we can use smoke or dust to make the light visible as it propagates between the slits and the screen. Assuming that the space in the figure is filled with smoke, what kind of light and dark pattern would you see as you look down? Draw the pattern on the figure by shading areas that would appear dark and leaving the white of the paper for areas that would appear bright.

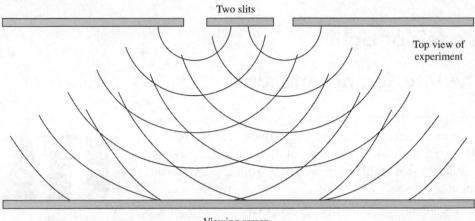

Two slits

Top view of experiment

Viewing screen

4. The figure shows the viewing screen in a double-slit experiment. Fringe C is the central maximum. For questions a–c, will the fringe spacing increase, decrease, or stay the same? Give an explanation for each.

a. The distance to the screen is increased.

b. The spacing between the slits is increased.

c. The wavelength of the light is increased.

d. Suppose the wavelength of the light is 500 nm. How much farther is it from the dot in the center of fringe E to the more distant slit than it is from the dot to the nearer slit?

33.3 The Diffraction Grating

5. The figure shows four slits in a diffraction grating. A set of Huygens wavelets is spreading out from each slit. Four wave paths, numbered 1 to 4, are shown leaving the slits at angle θ_1. The dashed lines are drawn perpendicular to the paths of the waves.

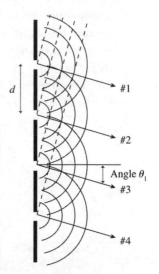

   a. Use a colored pencil or heavy shading to show *on the figure* the extra distance traveled by wave 1 that is not traveled by wave 2.

   b. How many extra wavelengths does wave 1 travel compared to wave 2? Explain how you can tell from the figure.

   c. How many extra wavelengths does wave 2 travel compared to wave 3?

   d. As these four waves combine at some large distance from the grating, will they interfere constructively, destructively, or in between? Explain.

   e. Suppose the wavelength of the light is doubled. (Imagine erasing every other wave front in the picture.) Would the interference at angle θ_1 then be constructive, destructive, or in between? Explain. Your explanation should be based on the figure, not on some equation.

   f. Suppose the slit spacing is doubled. (Imagine closing every other slit in the picture). Would the interference at angle θ_1 then be constructive, destructive, or in between? Again, base your explanation on the figure.

6. These are the same slits as in Exercise 5. Waves with the same wavelength are spreading out on the right side.

 a. Draw four paths, starting at the slits, at an angle θ_2 such that the wave along each path travels *two* wavelengths farther than the next wave. Also draw dashed lines at right angles to the travel direction. Your picture should look much like the figure of Exercise 5, but with the waves traveling at a different angle. Use a ruler!

 b. Do the same for four paths at angle $\theta_{1/2}$ such that each wave travels *one-half* wavelength farther than the next wave.

$\Delta r = 2\lambda$

$\Delta r = \frac{1}{2}\lambda$

7. This is the interference pattern on a viewing screen behind two slits. How would the pattern change if the two slits were replaced by 20 slits having the *same spacing d* between adjacent slits?

 a. Would the number of fringes on the screen increase, decrease, or stay the same?

 b. Would the fringe spacing increase, decrease, or stay the same?

 c. Would the width of each fringe increase, decrease, or stay the same?

 d. Would the brightness of each fringe increase, decrease, or stay the same?

33.4 Single-Slit Diffraction

33.5 A Closer Look at Diffraction

8. Plane waves of light are incident on two narrow, closely spaced slits. The graph shows the light intensity seen on a screen behind the slits.

   a. Draw a graph on the axes below to show the light intensity on the screen if the right slit is blocked, allowing light to go only through the left slit.

   b. Explain why the graph will look this way.

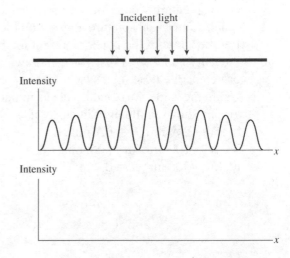

Incident light

Intensity

x

Intensity

x

9. This is the light intensity on a viewing screen behind a slit of width a. The light's wavelength is λ. Is $\lambda < a$, $\lambda = a$, $\lambda > a$, or is it not possible to tell? Explain.

10. This is the light intensity on a viewing screen behind a rectangular opening in a screen. Is the shape of the opening

 or ▯ or ▯ ?

Explain.

11. The graph shows the light intensity on a screen behind a 0.2-mm-wide slit illuminated by light with a 500 nm wavelength.

    a. Draw a *picture* in the box of how a photograph taken at this location would look. Use the same horizontal scale, so that your picture aligns with the graph above. Let the white of the paper represent the brightest intensity and the darkest you can draw with a pencil or pen be the least intensity.

    b. Using the same horizontal scale as in part a, draw graphs showing the light intensity if
       i. $\lambda = 250$ nm, $a = 0.2$ mm.
       ii. $\lambda = 1000$ nm, $a = 0.2$ mm.
       iii. $\lambda = 500$ nm, $a = 0.1$ mm.

Intensity $\lambda = 500$ nm $a = 0.2$ mm

Intensity $\lambda = 250$ nm $a = 0.2$ mm

Intensity $\lambda = 1000$ nm $a = 0.2$ mm

Intensity $\lambda = 500$ nm $a = 0.1$ mm

33.6 Circular-Aperture Diffraction

12. This is the light intensity on a viewing screen behind a circular aperture.

    a. If the wavelength of the light is increased, will the width of the central maximum increase, decrease, or stay the same? Explain.

    b. If the diameter of the aperture is increased, will the width of the central maximum increase, decrease, or stay the same? Explain.

    c. How will the screen appear if the aperture diameter is decreased to less than the wavelength of the light?

33.7 The Wave Model of Light

No exercises for this section

33.8 Interferometers

13. The figure shows a tube through which sound waves with $\lambda = 4$ cm travel from left to right. Each wave divides at the first junction and recombines at the second. The dots and triangles show the positions of the wave crests at $t = 0$ s—rather like a very simple wave front diagram.

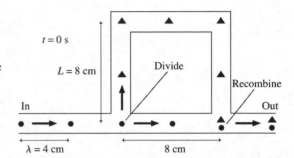

a. Do the recombined waves interfere constructively or destructively? Explain.

b. How much *extra* distance does the upper wave travel? _____

How many wavelengths is this extra distance? _____

c. Below are tubes with $L = 9$ cm and $L = 10$ cm. Use dots to show the wave crest positions at $t = 0$ s for the wave taking the lower path. Use triangles to show the wave crests at $t = 0$ s for the wave taking the upper path. The wavelength is $\lambda = 4$ cm. Assume that the first crest is at the left edge of the tube, as in the figure above.

d. How many *extra* wavelengths does the upper wave travel in the $L = 9$ cm tube? _____

What kind of interference does the $L = 9$ cm tube produce? _____

e. How many *extra* wavelengths does the upper wave travel in the $L = 10$ cm tube? _____

What kind of interference does the $L = 10$ cm tube produce? _____

14. A Michelson interferometer has been adjusted to produce a bright spot at the center of the interference pattern.

a. Suppose the wavelength of the light is halved. Is the center of the pattern now bright or dark, or is it not possible to say? Explain.

b. Suppose the wavelength of the light doubled to twice its original value. Is the center of the pattern now bright or dark, or is it not possible to say? Explain.

34 Ray Optics

Note: Please use a ruler or straight edge for drawing light rays.

34.1 The Ray Model of Light

1. If you turn on your car headlights during the day, the road ahead of you doesn't appear to get brighter. Why not?

2. a. Draw four or five rays from object 1 that allow A to see object 1.

   b. Draw four or five rays from object 2 that allow B to see object 2.

   c. What happens to the light where the rays cross in the center of the picture?

   1 ☆

   2 ☆

   B

   A

3. Are the following most reasonably described by the wave model of light, the ray model of light, either, or neither?

   a. The human eye _____

   b. A telescope _____

   c. Antireflection coatings on the telescope lenses _____

   d. Light passing through a 1-cm-diameter hole _____

   e. Light passing through a 0.1-mm-diameter hole _____

   f. An interferometer _____

   g. White light dispersed into a rainbow by a prism _____

   h. White light dispersed into a rainbow by a diffraction grating _____

4. A point source of light illuminates a slit in an opaque barrier.

   a. On the screen, sketch the pattern of light that you expect to see. Let the white of the paper represent light areas; shade dark areas. Mark any relevant dimensions.

   b. What will happen to the pattern of light on the screen if the slit width is reduced to 0.5 cm?

5. In each situation below, light passes through a 1-cm-diameter hole and is viewed on a screen. For each, sketch the pattern of light that you expect to see on the screen. Let the white of the paper represent light areas; shade dark areas.

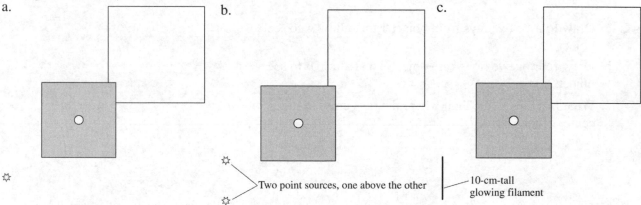

a.

b. Two point sources, one above the other

c. 10-cm-tall glowing filament

6. Light from a bulb passes through a pinhole. On the screen, sketch the pattern of light that you expect to see.

34.2 Reflection

7. a. Using a straight edge, draw five rays from the object that pass through points A to E after reflecting from the mirror. Make use of the grid to do this accurately.

   b. Extend the reflected rays behind the mirror.

   c. Show and label the image point.

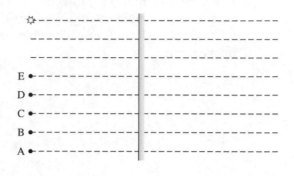

8. a. Draw *one* ray from the object that enters the eye after reflecting from the mirror.

   b. Is one ray sufficient to tell your eye/brain where the image is located?

   c. Use a different color pen or pencil to draw two more rays that enter the eye after reflecting. Then use the three rays to locate (and label) the image point.

   d. Do any of the rays that enter the eye actually pass through the image point?

9. You are looking at the image of a pencil in a mirror.

   a. What happens to the image you see if the top half of the mirror, down to the midpoint, is covered with a piece of cardboard? Explain.

— Midpoint

   b. What happens to the image you see if the bottom half of the mirror is covered with a piece of cardboard? Explain.

10. The two mirrors are perpendicular to each other.

   a. *Use a ruler* to draw a ray directly from the object to point A. Then draw two rays that strike the mirror about 3 mm (1/8 in) on either side of A. Draw the reflections of these three rays, making sure each obeys the law of reflection, then extend the reflections either forward or backward to locate an image point.

   b. Do the same for points B, C, and D.

   c. How many images are there, and where are they located?

34.3 Refraction

11. Draw seven rays from the object that refract after passing through the seven dots on the boundary.

a.

b.

c.

12. Complete the trajectories of these three rays through material 2 and back into material 1. Assume $n_2 < n_1$.

13. The figure shows six conceivable trajectories of light rays leaving an object. Which, if any, of these trajectories are impossible? For each that is possible, what are the requirements of the index of refraction n_2?

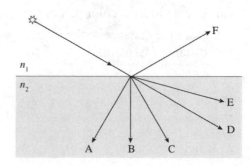

Impossible ....

Requires $n_2 > n_1$ ....

Requires $n_2 = n_1$ ....

Requires $n_2 < n_1$ ....

Possible for any n_2 ....

14. Complete the ray trajectories through the two prisms shown below.

a.

b.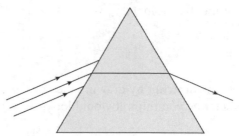

15. Draw the trajectories of seven rays that leave the object heading toward the seven dots on the boundary. Assume $n_2 < n_1$ and $\theta_c = 45°$.

34.4 Image Formation by Refraction at a Plane Surface

16. a. Draw rays that refract after passing through points B, C, and D. Assume $n_2 > n_1$.

    b. Use dashed lines to extend these rays backward into medium 1. Locate and label the image point.

    c. Now draw the rays that refract at A and E.

    d. Use a different color pen or pencil to draw three rays from the object that enter the eye.

    e. Does the distance to the object *appear* to be larger than, smaller than, or the same as the true distance? Explain.

17. A thermometer is partially submerged in an aquarium. The underwater part of the thermometer is not shown.

    a. As you look at the thermometer, does the underwater part appear to be closer than, farther than, or the same distance as the top of the thermometer?

    b. Complete the drawing by drawing the bottom of the thermometer as you think it would look.

34.5 Thin Lenses: Ray Tracing

18. a. Continue these rays through the lens and out the right side.

    b. Is the point where the rays converge the same as the focal point of the lens? Or different? Explain.

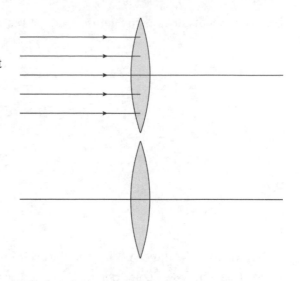

    c. Place a point source of light at the place where the rays converged in part b. Draw several rays heading left, toward the lens. Continue the rays through the lens and out the left side.

    d. Do these rays converge? If so, where?

19. The top two figures show test data for a lens. The third figure shows a point source near this lens and four rays heading toward the lens.

    a. For which of these rays do you know, from the test data, its direction after passing through the lens?

    b. Draw the rays you identified in part a as they pass through the lens and out the other side.

    c. Use a different color pen or pencil to draw the trajectories of the other rays.

    d. Label the image point. What kind of image is this?

    e. The fourth figure shows a second point source. Use ray tracing to locate its image point.

    f. The fifth figure shows an extended object. Have you learned enough to locate its image? If so, draw it.

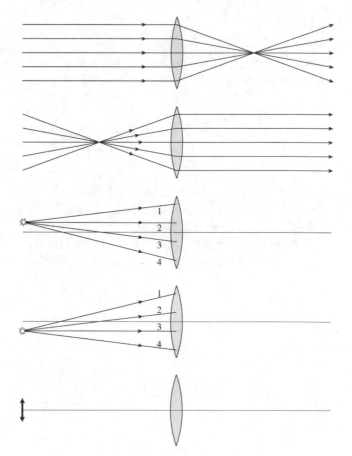

20. An object is near a lens whose focal points are marked with dots.

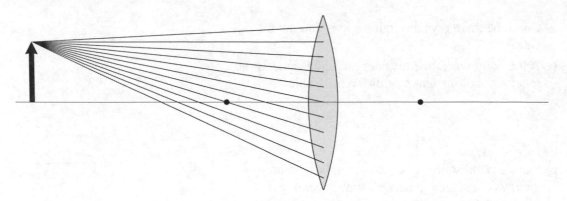

    a. Identify the three special rays and continue them through the lens.
    b. Use a different color pen or pencil to draw the trajectories of the other rays.

21. a. Consider *one* point on an object near a lens. What is the minimum number of rays needed to locate its image point?

    b. For each point on the object, how many rays from this point actually strike the lens and refract to the image point?

22. An object and lens are positioned to form a well-focused, inverted image on a viewing screen. Then a piece of cardboard is lowered just in front of the lens to cover the *top half* of the lens. Describe what happens to the image on the screen. What will you see when the cardboard is in place?

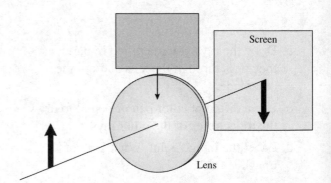

23. An object is near a lens whose focal points are shown.

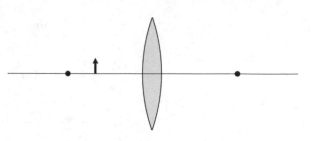

   a. Use ray tracing to locate the image of this object.
   b. Is the image upright or inverted?

   c. Is the image height larger or smaller than the object height?

   d. Is this a real or a virtual image? Explain how you can tell.

24. The top two figures show test data for a lens. The third figure shows a point source near this lens and four rays heading toward the lens.

   a. For which of these rays do you know, from the test data, its direction after passing through the lens?

   b. Draw the rays you identified in part a as they pass through the lens and out the other side.

   c. Use a different color pen or pencil to draw the trajectories of the other rays.

   d. Find and label the image point. What kind of image is this?

34.6 Thin Lenses: Refraction Theory

25. Materials 1 and 2 are separated by a spherical surface. For each part:
    i.   Draw the normal to the surface at the seven dots on the boundary.
    ii. Draw the trajectories of seven rays from the object that pass through the seven dots.
    iii. Trace the refracted rays either forward to a point where they converge or backward to a point from which they appear to diverge.

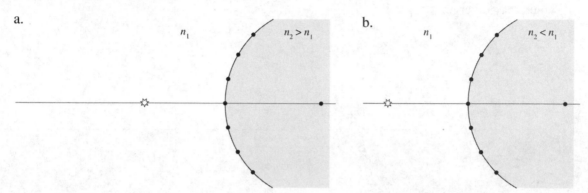

26. A converging lens forms a real image. Suppose the object is moved farther from the lens. Does the image move toward or away from the lens? Explain.

27. A converging lens forms a virtual image. Suppose the object is moved closer to the lens. Does the image move toward or away from the lens? Explain.

34.7 Image Formation with Spherical Mirrors

28. Two spherical mirrors are shown. The center of each is marked. For each:
    i.   Draw the normal to the surface at the seven dots on the boundary.
    ii. Draw the trajectories of seven rays that strike the mirror surface at the dots and then reflect, obeying the law of reflection.
    iii. Trace the reflected rays either forward to a point where they converge or backward to a point from which they diverge.

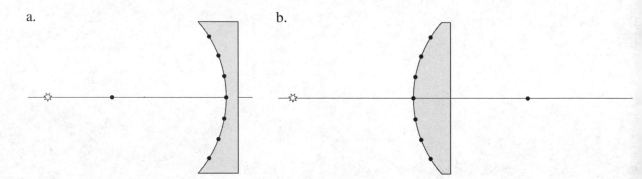

a.   b.

29. An object is placed near a spherical mirror whose focal point is marked.

   a. Identify the three special rays and show their reflections.

   b. Use a different color pen or pencil to draw the trajectories of the other rays.

35 Optical Instruments

35.1 Lenses in Combination

35.2 The Camera

1. Use ray tracing to locate the final image of the following two-lens system.

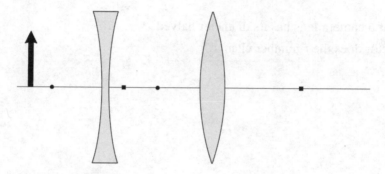

2. Can you tell what's inside the box? Draw one possible combination of lenses inside the box, then show the rays passing through the box.

3. Two converging lenses, whose focal points are marked, are placed in front of an object.

   a. Suppose lens 2 is moved a little to the left. Does the final image of this two-lens system move to the left, move to the right, or not move? Explain.

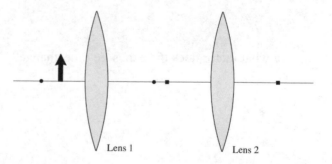

   b. With lens 2 in its original position, suppose lens 1 is moved a little to the left. Does the final image of this two-lens system move to the left, move to the right, or not move? Explain.

4. A photographer focuses his camera on an object. Suppose the object moves closer to the camera. To refocus, should the camera lens move closer to or farther from the detector? Explain.

5. The aperture of a camera lens has its diameter halved.

   a. By what factor does the *f*-number change?

   b. By what factor does the focal length change?

   c. By what factor does the exposure time change?

35.3 Vision

6. Two lost students wish to start a fire to keep warm while they wait to be rescued. One student is hyperopic, the other myopic. Which, if either, could use his glasses to focus the sun's rays to an intense bright point of light? Explain.

7. Suppose you wanted special glasses designed to let you see underwater, without a face mask. Should the glasses use a converging or diverging lens? Explain.

35.4 Optical Systems That Magnify

8. a. The angular magnification of a lens is not sufficient. To double the angular magnification, do you want a lens with twice the focal length or half the focal length? Explain.

   b. Does doubling the angular magnification also double the lateral magnification? Explain.

9. For a telescope, increasing the focal length of the objective increases the overall magnification. For a microscope, increasing the focal length of the objective decreases the overall magnification. Why are they different?

35.5 Color and Dispersion

10. A beam of white light from a flashlight passes through a red piece of plastic.

    a. What is the color of the light that emerges from the plastic?

    b. Is the emerging light as intense as, more intense than, or less intense than the white light?

    c. The light then passes through a blue piece of plastic. Describe the color and intensity of the light that emerges.

11. Suppose you looked at the sky on a clear day through pieces of red and blue plastic oriented as shown. Describe the color and brightness of the light coming through sections 1, 2, and 3.

    Section 1:

    Section 2:

    Section 3:

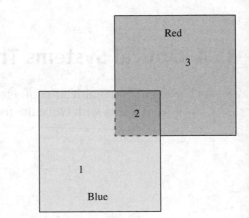

12. Sketch a plausible absorption spectrum for a patch of bright red paint.

13. Sketch a plausible absorption spectrum for a piece of green plastic.

35.6 The Resolution of Optical Instruments

14. A diffraction-limited lens can focus light to a 10-μm-diameter spot on a screen. Do the following actions make the spot diameter larger, smaller, or leave it unchanged?

    a. Decreasing the wavelength of the light: _____

    b. Decreasing the lens diameter: _____

    c. Decreasing the lens focal length: _____

    d. Decreasing the lens-to-screen distance: _____

15. An astronomer is trying to observe two distant stars. The stars are marginally resolved when she looks at them through a filter that passes green light near 550 nm. Which of the following actions would improve the resolution? Assume that the resolution is not limited by the atmosphere.

    a. Changing the filter to a different wavelength? If so, should she use a shorter or a longer wavelength?

    b. Using a telescope with an objective lens of the same diameter but a different focal length? If so, should she select a shorter or a longer focal length?

    c. Using a telescope with an objective lens of the same focal length but a different diameter? If so, should she select a larger or a smaller diameter?

    d. Using an eyepiece with a different magnification? If so, should she select an eyepiece with more or less magnification?

36 Relativity

36.1 Relativity: What's It All About?

36.2 Galilean Relativity

1. In which reference frame, S or S′, does the ball move faster?

2. Frame S′ moves relative to frame S as shown.

   a. A ball is at rest in frame S′. What are the speed and direction of the ball in frame S?

   b. A ball is at rest in frame S. What are the speed and direction of the ball in frame S′?

3. Frame S′ moves parallel to the *x*-axis of frame S.

   a. Is there a value of *v* for which the ball is at rest in S′? If so, what is *v*? If not, why not?

   b. Is there a value of *v* for which the ball has a minimum speed in S′? If so, what is *v*? If not, why not?

4. a. What are the speed and direction of each ball in a reference frame that moves with ball 1?

b. What are the speed and direction of each ball in a reference frame that moves with ball 2?

5. What are the speed and direction of each ball in a reference frame that moves to the right at 2 m/s?

36.3 Einstein's Principle of Relativity

6. A lighthouse beacon alerts ships to the danger of a rocky coastline.

   a. According to the lighthouse keeper, with what speed does the light leave the lighthouse?

   b. A boat is approaching the coastline at speed $0.5c$. According to the captain, with what speed is the light from the beacon approaching her boat?

7. As a rocket passes the earth at $0.75c$, it fires a laser perpendicular to its direction of travel.

   a. What is the speed of the laser beam relative to the rocket?

   b. What is the speed of the laser beam relative to the earth?

8. Teenagers Sam and Tom are playing chicken in their rockets. As seen from the earth, each is traveling at $0.95c$ as he approaches the other. Sam fires a laser beam toward Tom.

   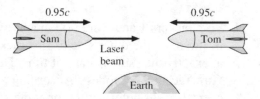

   a. What is the speed of the laser beam relative to Sam?

   b. What is the speed of the laser beam relative to Tom?

36.4 Events and Measurements

9. It is a bitter cold day at the South Pole, so cold that the speed of sound is only 300 m/s. The speed of light, as always, is 300 m/μs. A firecracker explodes 600 m away from you.

    a. How long after the explosion until you see the flash of light? _____

    b. How long after the explosion until you hear the sound? _____

    c. Suppose you see the flash at $t = 2.000\,002$ s. At what time was the explosion? _____

    d. What are the spacetime coordinates for the event "firecracker explodes"? Assume that you are at the origin and that the explosion takes place at a position on the positive x-axis.

10. Firecracker 1 is 300 m from you. Firecracker 2 is 600 m from you in the same direction. You see both explode at the same time. Define event 1 to be "firecracker 1 explodes" and event 2 to be "firecracker 2 explodes." Does event 1 occur before, after, or at the same time as event 2? Explain.

11. Firecrackers 1 and 2 are 600 m apart. You are standing exactly halfway between them. Your lab partner is 300 m on the other side of firecracker 1. You see two flashes of light, from the two explosions, at exactly the same instant of time. Define event 1 to be "firecracker 1 explodes" and event 2 to be "firecracker 2 explodes." According to your lab partner, based on measurements he or she makes, does event 1 occur before, after, or at the same time as event 2? Explain.

12. Two trees are 600 m apart. You are standing exactly halfway between them and your lab partner is at the base of tree 1. Lightning strikes both trees.

a. Your lab partner, based on measurements he or she makes, determines that the two lightning strikes were simultaneous. What did you see? Did you see the lightning hit tree 1 first, hit tree 2 first, or hit them both at the same instant of time? Explain.

b. Lightning strikes again. This time your lab partner sees both flashes of light at the same instant of time. What did you see? Did you see the lightning hit tree 1 first, hit tree 2 first, or hit them both at the same instant of time? Explain.

c. In the scenario of part b, were the lightning strikes simultaneous? Explain.

13. You are at the origin of a coordinate system containing clocks, but you're not sure if the clocks have been synchronized. The clocks have reflective faces, allowing you to read them by shining light on them. You flash a bright light at the origin at the instant your clock reads $t = 2.000\,000$ s.

a. At what time will you see the reflection of the light from a clock at $x = 3000$ m?

b. When you see the clock at $x = 3000$ m, it reads $2.000\,020$ s. Is the clock synchronized with your clock at the origin? Explain.

36.5 The Relativity of Simultaneity

14. Two supernovas, labeled L and R, occur on opposite sides of a galaxy, at equal distances from the center. The supernovas are seen at the same instant on a planet at rest in the center of the galaxy. A spaceship is entering the galaxy from the left at a speed of $0.999c$ relative to the galaxy.

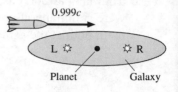

a. According to astronomers on the planet, were the two explosions simultaneous? Explain why.

b. Which supernova, L or R, does the spaceship crew *see* first? _____

c. Did the supernova that was *seen* first necessarily *happen* first in the rocket's frame? Explain.

d. Is "two light flashes reach the planet at the same instant" an event? To help you decide, could you arrange for something to happen only if two light flashes from opposite directions arrive at the same time? Explain.

If you answered Yes to part d, then the crew on the spaceship will also determine, from their measurements, that the light flashes reach the planet at the same instant. (Experimenters in different reference frames may disagree about when and where an event occurs, but they all agree that it *does* occur.)

e. The figure below shows the supernovas in the spaceship's reference frame with the *assumption* that the supernovas are simultaneous. The second half of the figure is a short time after the explosions. Draw two circular wave fronts on the second half of the figure to show the light from each supernova. Neither wave front has yet reached the planet. Be sure to consider:

- The points on which the wave fronts are centered.
- The wave speeds of each wave front.

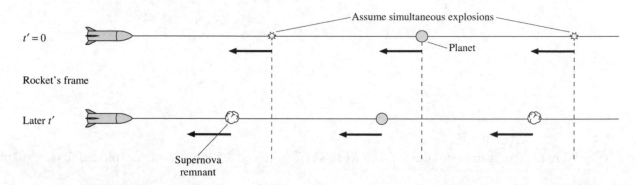

f. According to your diagram, are the two wave fronts going to reach the planet at the same instant of time? Why or why not?

g. Does your answer to part f conflict with your answer to part d? _____

If so, what different assumption could you make about the supernovas in the rocket's frame that would bring your wave-front diagram into agreement with your answer to part d?

h. So according to the spaceship crew, are the two supernovas simultaneous? If not, which happens first?

15. Peggy is standing at the center of her railroad car as it passes Ryan on the ground. Firecrackers attached to the ends of the car explode. A short time later, the flashes from the two explosions arrive at Peggy at the same time.

    a. Were the explosions simultaneous in Peggy's reference frame? If not, which exploded first? Explain.

    b. Were the explosions simultaneous in Ryan's reference frame? If not, which exploded first? Explain.

16. A rocket is traveling from left to right. At the instant it is halfway between two trees, lightning simultaneously (in the rocket's frame) hits both trees.

    a. Do the light flashes reach the rocket pilot simultaneously? If not, which reaches him first? Explain.

    b. A student was sitting on the ground halfway between the trees as the rocket passed overhead. According to the student, were the lightning strikes simultaneous? If not, which tree was hit first? Explain.

36.6 Time Dilation

17. Clocks C_1 and C_2 in frame S are synchronized. Clock C' moves at speed v relative to frame S. Clocks C' and C_1 read exactly the same as C' goes past. As C' passes C_2, is the time shown on C' earlier than, later than, or the same as the time shown on C_2? Explain.

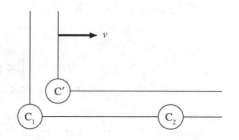

18. Your friend flies from Los Angeles to New York. She carries an accurate stopwatch with her to measure the flight time. You and your assistants on the ground also measure the flight time.

    a. Identify the two events associated with this measurement.

    b. Who, if anyone, measures the proper time? _____

    c. Who, if anyone, measures the shorter flight time? _____

19. You're passing a car on the highway. You want to know how much time is required to completely pass the car, from no overlap between the cars to no overlap between the cars. Call your car A and the car you are passing B.

    a. Specify two events that can be given spacetime coordinates. In describing the events, refer to cars A and B and to their front bumpers and rear bumpers.

    b. In either reference frame, is there *one* clock that is present at both events? _____

    c. Who, if anyone, measures the proper time between the events? _____

36.7 Length Contraction

20. Your friend flies from Los Angeles to New York. He determines the distance using the tried-and-true $d = vt$. You and your assistants on the ground also measure the distance, using meter sticks and surveying equipment.

    a. Who, if anyone, measures the proper length? _____

    b. Who, if anyone, measures the shorter distance? _____

21. Experimenters in B's reference frame measure $L_A = L_B$. Do experimenters in A's reference frame agree that A and B are the same length? If not, which do they find to be longer? Explain.

22. As a meter stick flies past you, you simultaneously measure the positions of both ends and determine that $L < 1$ m.

    a. To an experimenter in frame S', the meter stick's frame, did you make your two measurements simultaneously? If not, which end did you measure first? Explain.

       Hint: Review the reasoning about simultaneity that you used in Exercises 14–16.

    b. Can experimenters in frame S' give an explanation for why your measurement is < 1 m?

36.8 The Lorentz Transformations

23. A rocket travels at speed $0.5c$ relative to the earth.

    a. The rocket shoots a bullet in the forward direction at speed $0.5c$ relative to the rocket. Is the bullet's speed relative to the earth less than, greater than, or equal to c?

    b. The rocket shoots a second bullet in the backward direction at speed $0.5c$ relative to the rocket. In the earth's frame, is the bullet moving right, moving left, or at rest?

24. The rocket speeds are shown relative to the earth. Is the speed of A relative to B greater than, less than, or equal to $0.8c$?

    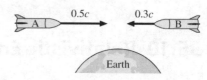

36.9 Relativistic Momentum

25. Particle A has half the mass and twice the speed of particle B. Is p_A less than, greater than, or equal to p_B? Explain.

26. Particle A has one-third the mass of particle B. The two particles have equal momenta. Is u_A less than, greater than, or equal to $3u_B$? Explain.

27. Event A occurs at spacetime coordinates (300 m, 2 μs).

a. Event B occurs at spacetime coordinates (1200 m, 6 μs). Could A possibly be the cause of B? Explain.

b. Event C occurs at spacetime coordinates (2400 m, 8 μs). Could A possibly be the cause of C? Explain.

36.10 Relativistic Energy

28. Can a particle of mass m have total energy less than mc^2? Explain.

29. Consider these 4 particles:

Particle	Rest energy	Total energy
1	A	A
2	B	$2B$
3	$2C$	$4C$
4	$3D$	$5D$

Rank in order, from largest to smallest, the particles' speeds u_1 to u_4.

Order:

Explanation:

37 The Foundations of Modern Physics

Note: Most of the exercises in this chapter are to give you practice reasoning using *macroscopic* evidence and observations to reason about *microscopic* processes.

37.2 The Emission and Absorption of Light

1. The figure shows the spectrum of a gas discharge tube.

   What color would the discharge appear to your eye? Explain.

2. Light is emitted from a discharge tube only if current flows through the gas. The light immediately ceases if the current is interrupted for any reason. What conclusions can you draw about the *mechanism* by which the light is emitted?

3. Spheres A and B are made of the same material and have equal diameters. The spectrum of thermal radiation from sphere A peaks at a wavelength one-third that at which B peaks. What is the ratio P_A/P_B of the power radiated by A to that radiated by B?

37.3 Cathode Rays and X Rays

37.4 The Discovery of the Electron

4. Over 100 years ago it was observed that a nearby source of x rays or radioactivity causes a charged electroscope to discharge. Replacing the air around the electroscope by a monatomic gas such as helium does not change the outcome. However, no discharge occurs if the electroscope is in a vacuum.

   a. Do the x rays or radioactivity affect the electroscope itself, or do they affect the environment around the electroscope? Use the observations to explain your reasoning.

   b. What is the *mechanism* by which the discharge occurs?

   c. What can you infer from these observations about the properties of atoms? Possibilities you might want to consider are the size of atoms, the mass of atoms, the divisibility of atoms, the structure of atoms, and perhaps other properties.

5. A magnetic field exists in the space between two parallel electrodes.

   a. Sketch the trajectory of the electron if the electrodes are uncharged.

   b. Use plus and minus symbols to show how the electrodes should be charged to achieve zero deflection of the electron.

6. Thomson observed deflection of the cathode-ray particles due to magnetic and electric fields, but there was no observed deflection due to gravity. Why not?

7. Cathode rays cause a green spot to appear on the glass at the front of a Crookes tube. A horizontal magnetic field deflects the spot up and down, but the spot remains the *same size*. It does not smear or spread out. This is an interesting observation.

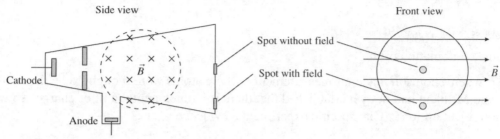

a. Assuming that cathode rays are negatively charged particles, what can you conclude about the charge-to-mass ratios of different cathode-ray particles?

b. What can you conclude about the speeds of different cathode-ray particles?

8. In the absence of fields, the green dot caused by the cathode rays appears in the center of the front face of the tube, as was shown in the figure of Exercise 7. Prior to Thomson, it was discovered that an electrode in the center of the front face collects a current of negative charge.

a. Can you conclude from this observation that cathode rays are negatively charged particles? Or is the evidence insufficient to draw this conclusion? Explain.

b. Why did Thomson carry out the experiment in which he placed the collecting electrode off to the side of the tube? What was the significance of this experiment?

9. Thomson found that $(q/m)_{\text{cathode rays}} \approx 1000(q/m)_{\text{H-ion}}$. (The charge-to-mass ratio of the hydrogen ion H^+ was known from electrolysis.) This suggests either:

    i. $q_{\text{cathode rays}} \approx 1000 q_{\text{H-ion}}$,

    ii. $m_{\text{cathode rays}} \approx \dfrac{1}{1000} m_{\text{H-ion}}$, or

    iii. Some combination of these.

    In 1897, when Thomson made his measurements, whole atoms were thought to be the smallest unit of matter and Millikan had not yet established that there is a fundamental unit of charge. So why did Thomson reject option i? Cite specific experimental evidence.

10. Thomson used cathodes of aluminum, iron, platinum, and other metals. He found the same value of $(q/m)_{\text{cathode rays}}$ for each. What conclusion can you draw from this observation?

37.5 The Fundamental Unit of Charge

11. The figure shows a positively charged oil drop held at rest between two electrodes.

   a. Draw a free-body diagram for the oil drop while it is at rest. Label all forces.

   b. If the electric field is turned off, the drop begins to fall very slowly at constant speed. Draw a free-body diagram of the drop as it falls. Label all forces.

12. Millikan observed both positive and negative oil drops.

   a. Why didn't all the oil drops have the same sign?

   b. Did the fact that he observed both positive and negative drops affect his conclusion that he had measured the "fundamental unit" of charge? Why or why not?

13. Charge measurements on six oil drops give the following values, all in some unknown units of charge.

Drop	q	Drop	q
1	+7.0	4	−28.0
2	+24.5	5	+14.0
3	−7.0	6	−10.5

In these units, what is the fundamental unit of charge? Explain.

37.6 The Discovery of the Nucleus

37.7 Into the Nucleus

14. A block of uranium emits alpha particles continuously. As Rutherford showed, each alpha particle has a charge and a mass.

    a. Does the continuous emission of these particles violate the law of conservation of mass? If not, why not?

    b. Does the continuous emission of these particles violate the law of conservation of charge? If not, why not?

15. A proton gains 50 eV of kinetic energy by moving through a potential difference. What is ΔV?

16. An electron, a proton, and a helium nucleus (He^{++}) accelerate through the 10 V potential difference between two capacitor plates. For each, what is the particle's increase in kinetic energy ΔK in electron volts?

    Electron $\Delta K =$ _____

    Proton $\Delta K =$ _____

    He^{++} $\Delta K =$ _____

17. Rutherford studied alpha particles using the same crossed-field technique that Thomson had invented to study cathode rays. Assuming that $v_{alpha} \approx v_{cathode\ ray}$ (which turns out to be true), would the deflection of an alpha particle by a magnetic field be larger than, smaller than, or the same as the deflection of a cathode-ray particle by the same field? Explain.

18. Once Thomson showed that atoms consist of very light negative electrons and a much more massive positive charge, why didn't physicists immediately consider a solar-system model of electrons orbiting a positive nucleus? Why would physicists in 1900 object to such a model?

19. Suppose you throw a small, hard rubber ball through a tree. The tree has many outer leaves, so you cannot see clearly into the tree. Most of the time your ball passes through the tree and comes out the other side with little or no deflection. On occasion, the ball emerges at a very large angle to your direction of throw. On rare occasions, it even comes straight back toward you. From these observations, what can you conclude about the structure of the tree? Be specific as to how you arrive at these conclusions *from the observations*.

20. Beryllium is the fourth element in the periodic table. Draw pictures similar to Figure 37.20 showing the structure of neutral Be, of Be$^+$, of Be^{++}, and of the negative ion Be$^-$.

Be   Be$^+$ Be^{++} Be$^-$

21. The element hydrogen has three isotopes. The most common has $A = 1$. A rare form of hydrogen (called *deuterium*) has $A = 2$. An unstable, radioactive form of hydrogen (called *tritium*) has $A = 3$. Draw pictures similar to Figure 37.23 showing the structure of these three isotopes. Show all the electrons, protons, and neutrons of each.

$A = 1$ $A = 2$ $A = 3$

22. Identify the element, the isotope, and the charge state. Give your answer in symbolic form, such as ^{4}He$^+$ or ^{8}Be$^-$.

a.

b.

38 Quantization

38.1 The Photoelectric Effect

38.2 Einstein's Explanation

1. a. A negatively charged electroscope can be discharged by shining an ultraviolet light on it. How does this happen?

UV light

   b. You might think that an ultraviolet light shining on an initially uncharged electroscope would cause the electroscope to become positively charged as photoelectrons are emitted. In fact, ultraviolet light has no noticeable effect on an uncharged electroscope. Why not?

2. In the photoelectric effect experiment, a current is measured while light is shining on the cathode. But this does not appear to be a complete circuit, so how can there be a current? Explain.

3. Draw the trajectories of several typical photoelectrons when (a) $\Delta V = V_{anode} - V_{cathode} > 0$, (b) $-V_{stop} < \Delta V < 0$, and (c) $\Delta V < -V_{stop}$.

a.   b.   c.

4. Why are only electrons emitted in the photoelectric effect experiment? Why are positive ions never emitted?

5. The work function of a metal measures:
   i.   The kinetic energy of the electrons in the metal.
   ii. How tightly the electrons are bound within the metal.
   iii. The amount of work done by the metal when it expands.
   Which of these (perhaps more than one) are correct? Explain.

6. The figure shows the typical photoelectric behavior of a metal as the anode-cathode potential difference ΔV is varied.

   a. Why do the curves become horizontal for $\Delta V > 0$ V? Shouldn't the current increase as the potential difference increases? Explain.

   b. Why doesn't the current immediately drop to zero for $\Delta V < 0$ V? Shouldn't $\Delta V < 0$ V prevent the electrons from reaching the anode? Explain.

c. The current is zero for $\Delta V < -V_{stop}$. Where do the photoelectrons go? Are no photoelectrons emitted if $\Delta V < -V_{stop}$? Or if they are, why is there no current? Explain.

7. a. What is the significance of V_{stop}? That is, what have you learned if you measure V_{stop}?

   Note: Don't say that "$-V_{stop}$ is the potential that causes the current to stop." That is merely the definition of V_{stop}. It doesn't say what the significance of V_{stop} is.

   b. Why is it surprising that V_{stop} is independent of the light intensity? What would you *expect* V_{stop} to do as the intensity increases? Explain.

   c. If the wavelength of the light in a photoelectric effect experiment is increased, does V_{stop} increase, decrease, or stay the same? Explain.

8. Metal 1 has a larger work function than metal 2. Both are illuminated with the same short-wavelength ultraviolet light. Do photoelectrons from metal 1 have a higher speed, a lower speed, or the same speed as photoelectrons from metal 2? Explain.

9. Consider the following characteristics of the photoelectric effect:

   i.   The generation of photoelectrons.

   ii. The existence of a threshold frequency.

   iii. The photoelectric current increases with increasing light intensity.

   iv. The photoelectric current is independent of ΔV for $\Delta V > 0$.

   v.   The photoelectric current decreases slowly as ΔV becomes more negative.

   vi. The stopping potential is independent of the light intensity.

   vii. The photoelectron current appears instantly when the light is turned on.

   Which of these *cannot* be explained by classical physics?

10. The figure shows a current-versus-potential difference graph for a photoelectric effect experiment. On the figure, draw and label graphs for the following three situations:

    i.   The light intensity is increased.

    ii. The light frequency is increased.

    iii. The cathode work function is increased.

    In each case, no other parameters of the experiment are changed.

11. The figure shows a current-versus-frequency graph for a photoelectric effect experiment with $\Delta V > 0$. On the figure, draw and label graphs for the following three situations:

    i.   The light intensity is increased.

    ii. The anode-cathode potential difference is increased.

    iii. The cathode work function is increased.

    In each case, no other parameters of the experiment are changed.

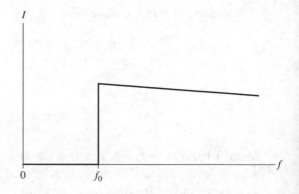

12. The figure shows a stopping potential-versus-frequency graph for a photoelectric effect experiment. On the figure, draw and label graphs for the following two situations:

    i.   The light intensity is increased.

    ii. The cathode work function is increased.

    In each case, no other parameters of the experiment are changed.

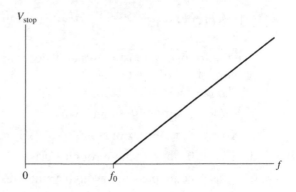

13. In a photoelectric effect experiment, the frequency of the light is increased while the intensity is held constant. As a result:

    i.   There are more photoelectrons.   iii. Both i and ii.

    ii. The photoelectrons are faster.   iv. Neither i nor ii.

    Explain your choice.

14. In a photoelectric effect experiment, the intensity of the light is increased while the frequency is held constant. As a result:

    i.   There are more photoelectrons.   iii. Both i and ii.

    ii. The photoelectrons are faster.   iv. Neither i nor ii.

    Explain your choice.

15. A gold cathode (work function $= 5.1$ eV) is illuminated with light of wavelength 250 nm. It is found that the photoelectron current is zero when $\Delta V = 0$ V. Would the current change if:

    a. The intensity is doubled?

    b. The anode-cathode potential difference is increased to $\Delta V = 5.5$ V?

    c. The cathode is changed to aluminum (work function $= 4.3$ eV)?

38.3 Photons

16. Are the following most reasonably described by the wave model of light, the ray model of light, or the photon model of light?

    a. The photoelectric effect _____

    b. Sunlight coming through a window _____

    c. X-ray scattering from atoms _____

    d. A laser beam passing through a prism _____

    e. A laser beam passing through a small hole _____

    f. Extremely weak light passing through a small hole _____

17. The intensity of a beam of light is increased but the light's frequency is unchanged. As a result:

    i. The photons travel faster.   iii. The photons are larger.

    ii. Each photon has more energy.   iv. There are more photons per second.

    Which of these (perhaps more than one) are true? Explain.

18. The frequency of a beam of light is increased but the light's intensity is unchanged. As a result:

    i. The photons travel faster.   iii. There are fewer photons per second.

    ii. Each photon has more energy.   iv. There are more photons per second.

    Which of these (perhaps more than one) are true? Explain.

19. Photons are sometimes represented by pictures like this. Just what is this a picture of? Explain what this "graph" shows. _____

20. Light of wavelength $\lambda = 1\mu$m is emitted from point A. A photon is detected 5 μm away at point B. On the figure, draw the trajectory that a photon follows between points A and B.

38.4 Matter Waves and Energy Quantization

21. Electron 1 is accelerated from rest through a potential difference of 100 V. Electron 2 is accelerated from rest through a potential difference of 200 V. Afterward, which electron has the larger de Broglie wavelength? Explain.

22. An electron and a proton are each accelerated from rest through a potential difference of 100 V. Afterward, which particle has the larger de Broglie wavelength? Explain.

23. Neutron beam 1 has a temperature of 500 K. Neutron beam 2 has a temperature of 5 K. Which neutrons have the larger de Broglie wavelength? Explain.

24. A neutron is shot straight up in vacuum with an initial speed of 100 m/s. As it rises, does its de Broglie wavelength increase, decrease, or not change? Explain.

25. Double-slit interference of electrons occurs because:
    i.   The electrons passing through the two slits repel each other.
    ii. Electrons collide with each other behind the slits.
    iii. Electrons collide with the edges of the slits.
    iv. Each electron goes through both slits.
    v.   The energy of the electrons is quantized.
    vi. Only certain wavelengths of the electrons fit through the slits.
      Which of these (perhaps more than one) are correct? Explain.

26. Can an electron with a de Broglie wavelength of 2 μm pass through a slit that is 1 μm wide?

27. The figure shows the standing de Broglie wave of a particle in a box.

    a. What is the quantum number? _____

    b. Can you determine from this picture whether the "classical" particle is moving to the right or to the left? If so, which is it? If not, why not?

28. A particle in a box of length L_a has $E_1 = 2$ eV. The same particle in a box of length L_b has $E_2 = 50$ eV. What is the ratio L_a/L_b?

29. The figure shows a standing de Broglie wave.

    a. Does this standing wave represent a particle that travels back and forth between the boundaries with a constant speed or a changing speed? Explain.

    b. If the speed is changing, at which end is the particle moving faster and at which end is it moving slower?

38.5 Bohr's Model of Atomic Quantization

38.6 The Bohr Hydrogen Atom

38.7 The Hydrogen Spectrum

30. J. J. Thomson studied the ionization of atoms in collisions with electrons. He accelerated electrons through a potential difference, shot them into a gas of atoms, then used a mass spectrometer to detect any ions produced in the collisions. By using different gases, he found that he could produce singly ionized atoms of all the elements that he tried. When he used higher accelerating voltages, he was able to produce doubly ionized atoms of all elements *except* hydrogen.

    a. Why did Thomson have to use higher accelerating voltages to detect doubly ionized atoms than to detect singly ionized atoms?

    b. What conclusion or conclusions about hydrogen atoms can you draw from these observations? Be specific as to how your conclusions are related to the observations.

31. If an electron is in a *stationary state* of an atom, is the electron at rest? If not, what does the term mean?

32. The diagram shows the energy level diagram of element X.

    a. What is the ionization energy of element X?

    $E = 0$ eV ‑ ‑ ‑ ‑ ‑ ‑ ‑ ‑ ‑ ‑ ‑ ‑
    -1 eV ——————— $n = 3$
    |1240 nm
    -2 eV ——————— $n = 2$

    -4 eV ——————— $n = 1$

    b. An atom in the ground state absorbs a photon, then emits a photon with a wavelength of 1240 nm. What conclusion can you draw about the energy of the photon that was absorbed?

    c. An atom in the ground state has a collision with an electron, then emits a photon with a wavelength of 1240 nm. What conclusion can you draw about the initial kinetic energy of the electron?

33. The figure shows a hydrogen atom, with an electron orbiting a proton.

    a. What force or forces act on the electron?

    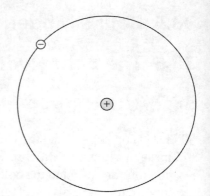

    b. On the figure, draw the electron's velocity, acceleration, and
       force vectors.

34. Bohr did not include the gravitational force in his analysis of the
    hydrogen atom. Is this one of the reasons that his model of the hydrogen atom had only limited
    success? Explain.

35. a. The stationary state of hydrogen shown on the left has quantum number $n =$ _____

    b. On the right, draw the stationary state of the $n - 1$ state.

36. The $n = 3$ state of hydrogen has $E_3 = -1.51$ eV.

    a. Why is the energy negative?

    b. What is the physical significance of the specific number 1.51 eV?

37. Why is there no stationary state of hydrogen with $E = -9$ eV?

38. Draw and label an energy level diagram for hydrogen. On it, show all the transitions by which an electron in the $n = 4$ state could emit a photon.

39. a. Which states of a hydrogen atom can be excited by collision with an electron with kinetic energy $K = 12.5$ eV? Explain.

    b. After the collision, the electron:

       i. Bounces off with $K > 12.5$ eV.

       ii. Bounces off with $K = 12.5$ eV.

       iii. Bounces off with $K < 12.5$ eV.

       iv. Has been absorbed.

       Explain your choice.

    c. After the collision, the atom emits a photon. List all the possible $n \rightarrow m$ transitions that might occur as a result of this collision.

40. The longest wavelength in the Balmer series is 656 nm.

    a. What transition is this?

    b. If light of this wavelength shines on a container of hydrogen atoms, will the light be absorbed? Why or why not?

41. a. How many electrons, protons, and neutrons are in a hydrogen-like ^{12}C ion?

    b. Draw a picture of a hydrogen-like ^{12}C ion, showing all the particles you identified in part a.

39 Wave Functions and Uncertainty

39.1 Waves, Particles, and the Double-Slit Experiment

39.2 Connecting the Wave and Photon Views

1. An experiment has three possible outcomes: A, B, and C. The probability of A is 20% and the probability of B is 30%.

   a. What is the probability that the outcome is either A or B? _____

   b. What is the probability that the outcome is C? _____

   c. What is the probability that the outcome is not B? _____

   d. Out of 1000 trials, how many do you *expect* to have the outcome A? _____

   e. If 1000 trials yielded 210 A's, would you be concerned that the stated probabilities are in error? Why or why not?

2. a. The graph shows the linear mass density of a string whose density is a function of position. Which end of the string (left or right) is more massive?

   Linear mass density
   μ (g/m)

   b. What is the mass of a 1-mm-long segment of string at:

   $x = 0.25$m   mass = _____
   $x = 0.50$ m   mass = _____
   $x = 0.75$ m   mass = _____

   c. This graph shows the probability density for photons to be detected on the *x*-axis. At which end of the $0 \text{ m} \le x \le 1 \text{ m}$ interval ($x \approx 0$ m or $x \approx 1$ m) is a photon more likely to be detected?

   Probability density
   P (m^{-1})

   d. 1,000,000 photons are detected. What is the expected number of photons in a 1-mm-wide interval at:

   $x = 0.25$ m   # photons = _____
   $x = 0.50$ m   # photons = _____
   $x = 0.75$ m   # photons = _____

3. A classical ball bounces back and forth between two rigid walls with no loss of speed. After a long time has passed, the ball's position is measured at a random time.

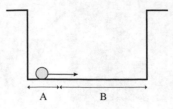

a. Is the probability Prob(in A) that the ball is found in section A larger than, smaller than, or the same as the probability Prob(in B) of being found in section B? Explain.

b. Is the probability density in section A larger than, smaller than, or the same as the probability density in section B? Explain.

4. Light passes through a single narrow slit. The top graph in the figure shows the measured light intensity $I(x)$ on a screen behind the slit.

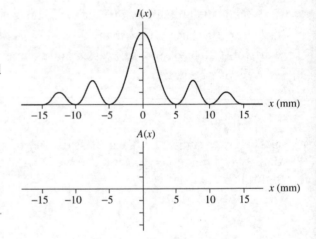

a. Graph the amplitude function $A(x)$. Keep in mind that the amplitude function *oscillates* between positive and negative values.

b. In the space below your amplitude graph, draw a dot pattern showing the likely distribution of the first ≈ 100 photons.

c. What are the most probable value or values of x for detecting a photon?

d. What are the least probable value or values of x for detecting a photon?

Photon distribution

e. 1,000,000 photons are detected. Make a rough estimate of the number that are detected in the interval -5 mm $< x < +5$ mm. Explain your reasoning.

39.3 The Wave Function

5. The figure shows the wave function of electrons arriving on a screen along the x-axis.

    a. Draw a graph of $|\psi(x)|^2$.

    b. Draw a dot pattern showing the likely distribution of electrons on the screen.

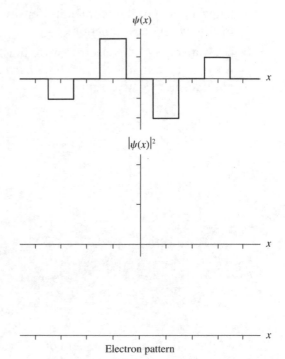

6. The figure shows the wave function of electrons arriving on a screen along the x-axis.

    a. Draw a graph of $|\psi(x)|^2$.

    b. Draw a dot pattern showing the likely distribution of electrons on the screen.

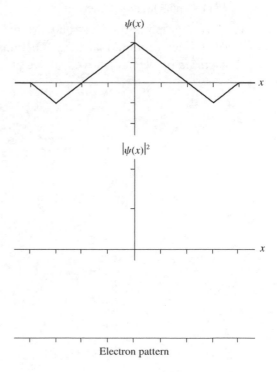

7. The figure shows the dot pattern of electrons landing on a screen.

   a. Draw a graph of the probability density $P(x) = |\psi(x)|^2$ as a function of x.

   b. Draw a graph of the electron wave function $\psi(x)$. Keep in mind that wave functions *oscillate* between positive and negative values.

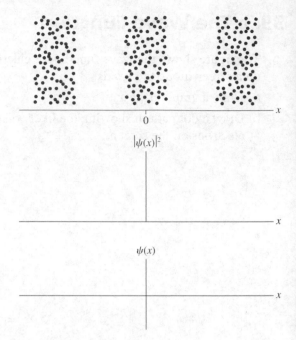

8. The figure shows the dot pattern of electrons landing on a screen.

   a. Draw a graph of the probability density $P(x) = |\psi(x)|^2$ as a function of x.

   b. Draw a graph of the electron wave function $\psi(x)$. Keep in mind that wave functions *oscillate* between positive and negative values.

39.4 Normalization

9. The figure shows a graph of $|\psi(x)|^2$.

    a. What value of the constant a makes this a normalized wave function? Explain.

    b. What is Prob(in 0.001 cm at $x = 0.25$ cm)?

    c. What is Prob(0 cm $\le x \le 0.5$ cm)?

    d. Draw a graph of the wave function $\psi(x)$. Add an appropriate vertical scale to your graph. (Think carefully about the proper shape.)

10. The figure shows the graph of a wave function $\psi(x)$.

    a. Draw the graph of $|\psi(x)|^2$. Add an appropriate vertical scale to your graph.

    b. Is this a normalized wave function? Explain.

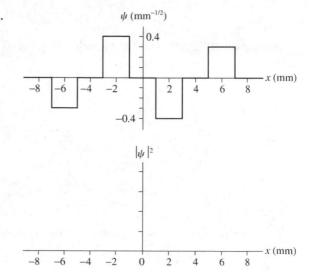

    c. What is the value of: Prob(5 mm $\le x \le 10$ mm)? _____

    Prob(-5 mm $\le x \le 5$ mm)? _____

    Prob(-10 mm $\le x \le 10$ mm)? _____

39.5 Wave Packets

39.6 The Heisenberg Uncertainty Principle

11. Approximately what range of frequencies must be superimposed to create this wave packet?

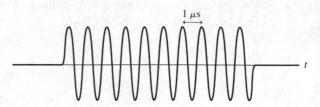

12. A telecommunication system uses pulses with a duration of 5 ns.

    a. About how many such pulses can be transmitted per second?

    b. What is the minimum bandwidth needed to transmit these pulses?

13. Wave packets are shown for particles 1, 2, and 3.

    Particle 1   Particle 2   Particle 3

    Which particle can have its velocity known most precisely? Explain.

40 One-Dimensional Quantum Mechanics

40.1 The Schrödinger Equation

40.2 Solving the Schrödinger Equation

1. The graphs of several functions are shown below. For each, would this function be an acceptable wave function in quantum mechanics? If not, what is wrong with it?

a.

b.

c.

d.

e.

f.

40.3 A Particle in a Rigid Box: Energies and Wave Functions

40.4 A Particle in a Rigid Box: Interpreting the Solution

40.5 The Correspondence Principle

2. Use the properties of waves to explain why a particle in a rigid box can have only certain energies but not others.

3. The graph shows the energy levels and wave functions of the $n = 2$ and $n = 4$ states of a particle in a box.

   a. In which state does the particle have more energy?

   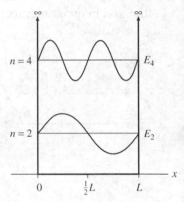

   b. In which state are you more likely to find the particle at $x = \frac{3}{4}L$? Explain.

4. Can an atom be at rest in the center of a microscopic box? If so, describe a procedure by which you could arrange this. If not, why not?

5. A classical mass on a spring oscillates in simple-harmonic motion between $x = -A$ and $x = +A$. The equilibrium position is $x = 0$. One thousand pictures of the mass are taken at random times over many thousands of oscillations. Where will the mass be in most of the photos? Explain.

40.6 Finite Potential Wells

6. A particle in a potential well is in the $n = 6$ quantum state. How many peaks are there in the probability density $P(x) = |\psi(x)|^2$?

7. What is the quantum number for this particle in a finite potential well? How can you tell?

8. The diagram shows a neutron occupying one energy level in a nucleus.

    a. How many possibilities are there for the emission of a gamma-ray photon?

    b. What is the energy, in MeV, of each gamma-ray photon that could be emitted?

9. Rank in order, from largest to smallest, the penetration distances η_1 to η_3 of the electron wave functions corresponding to these three energy levels.

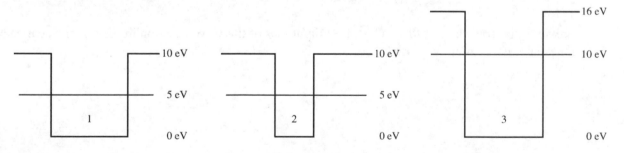

Order:

Explanation:

40.7 Wave-Function Shapes

10. Energy E is an allowed energy for a particle in this potential well. Three positions—A, B, and C—are marked on the x-axis.

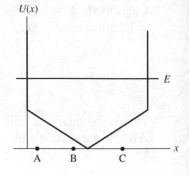

   a. Rank in order, from largest to smallest, the classical kinetic energies of the particle at these three positions.

   b. Rank in order, from largest to smallest, the particle's de Broglie wavelengths at these three positions.

   c. Rank in order, from largest to smallest, the amount of time a classical particle spends traversing an interval of width δx at each of these three points.

   d. Rank in order, from largest to smallest, the spacings between the zeros of the wave function in the regions near each of these points. Assume that $n \gg 1$ and that the wave function oscillates many times between the two walls of the potential energy. Explain your ranking.

   e. Rank in order, from largest to smallest, the amplitudes of the wave function in the regions near each of these points. Explain your ranking.

11. The figure shows the $n = 4$ and the $n = 8$ energy levels of a particle in a potential well. Based on your answers to Exercise 10, sketch plausible $n = 4$ and $n = 8$ wave functions.

12. The figure shows a solution of the Schrödinger equation.

a. What is the value of the quantum number n? Explain how you know.

b. What is the probability of finding the particle at $x < 0$?

What is the probability of finding the particle at $x > L$?

c. At what value or values of x is the particle most likely to be found?

d. In what general regions is the de Broglie wavelength the shortest?

In what general region is the de Broglie wavelength the longest?

e. In what general regions would a classical particle move the fastest?

In what general region would a classical particle move the slowest?

f. In what regions would a classical particle have maximum kinetic energy?

In what region would a classical particle have minimum kinetic energy?

g. Now, on the lower set of axes, sketch a plausible potential well in which the particle is confined.

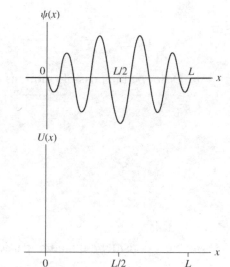

13. The figure shows the $n = 2$ and $n = 8$ energy
    levels of a particle in a potential well.

    a. Describe how a classical particle with energy E_8 would
       move. In particular, specify where it speeds up, where it
       slows down, where the turning points are, and which
       regions are forbidden to the particle.

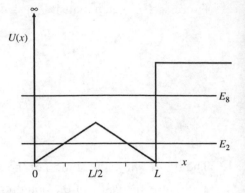

    b. Describe the possible motions of a classical particle with energy E_2.

    c. Now consider a quantum particle with energy E_8. What is the value of the wave function at $x = 0$?

       Can you determine the value of the wave function at $x = L$? If so, what is it? If not, why not?

       Sketch—on the E_8 energy level in the figure—a plausible wave function for energy E_8.

    d. Suppose the quantum particle has energy E_2. In what region or regions that would be forbidden to a
       classical particle might you find a quantum particle?

       Sketch—on the E_2 energy level—a plausible wave function for energy E_2.

40.8 The Quantum Harmonic Oscillator

40.9 More Quantum Models

14. A particle's wave function is shown at the right.

    a. What is the quantum number? _____

    b. In which potential well below is the particle confined? _____

15. The figure shows the $n = 2$ energy level and wave function for a quantum harmonic oscillator.

    a. Can you determine, from the figure, the amplitude of the wave function? If so, what is it? If not, why not?

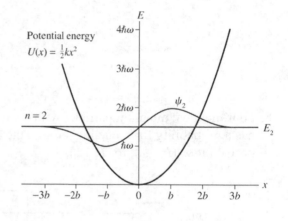

    b. On the figure, identify and label the classical turning points for a particle with energy E_2.

    c. On the figure, shade the area between the wave function and the E_2 line to show where the wave function penetrates into a classically forbidden region.

16. The figure shows two possible wave functions for an electron in a linear triatomic molecule. Which of these is a bonding orbital and which is an antibonding orbital? Explain how you can distinguish them.

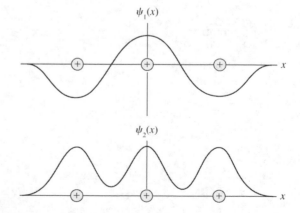

40.10 Quantum-Mechanical Tunneling

17. For the potential-energy functions shown below, spanning width L, can a particle on the left side of the potential well ($x < 0.45L$) with energy E tunnel into the right side of the potential well ($x > 0.55L$)? If not, why not?

a.   b.   c.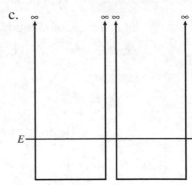

18. Four identical quantum particles with energy E approach a potential energy barrier from the left. Each has a probability for tunneling through the barrier. Rank in order, from largest to smallest, the tunneling probabilities $(P_{tunnel})_1$ to $(P_{tunnel})_4$.

Order:

Explanation:

41 Atomic Physics

41.1 The Hydrogen Atom: Angular Momentum and Energy

41.2 The Hydrogen Atom: Wave Functions and Probabilities

1. List all possible states of a hydrogen atom that have $E = -1.51$ eV.

$$n \qquad\qquad l \qquad\qquad m$$

2. What are the n and l values of the following states of a hydrogen atom?

   State $= 4d$ $n =$ _____ $l =$ _____

   State $= 5f$ $n =$ _____ $l =$ _____

   State $= 6s$ $n =$ _____ $l =$ _____

3. How would you label the hydrogen-atom states with the following quantum numbers?

   $(n, l, m) = (4, 3, 0)$ Label $=$ _____

   $(n, l, m) = (3, 2, 1)$ Label $=$ _____

   $(n, l, m) = (3, 2, -1)$ Label $=$ _____

4. Consider the two hydrogen-atom states $5d$ and $4f$. Which has the higher energy? Explain.

5. a. As a multiple of $\hbar$, what is the angular momentum of a d electron? _____

   b. What is the *maximum* z-component of angular momentum of a d electron? _____

   c. Is $(L_z)_{max}$ greater than, less than, or equal to L? _____

   d. What is the significance of your answer to part c?

6. Draw a picture similar to Figure 41.2 showing all possible orientations of the angular momentum vector for a p electron.

7. What is the difference between the *probability density* and the *radial probability density*?

8. Consider a 2s electron, as portrayed in Figures 41.4 and 41.5. In your own words, describe how these figures suggest a "shell structure" of electrons around the nucleus.

41.3 The Electron's Spin

9. The figure shows a spinning ball of negative charge. Does the magnetic moment of this spinning charge point up, point down, or point to the right? Explain.

10. A bar magnet is moving to the right through a nonuniform magnetic field. The field is weaker toward the bottom of the page and stronger toward the top of the page.

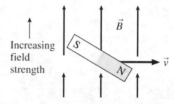

    a. Is there a net force on the magnet? If so, in which direction? Explain.

    b. Will the magnet be deflected by the field? If so, in which direction?

    c. Would the magnet be deflected by a *uniform* magnetic field? Explain why or why not.

11. The figure shows the outcome of a Stern-Gerlach experiment with atoms of element X.

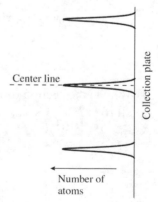

    a. Do the peaks in a Stern-Gerlach experiment represent different values of the atom's total angular momentum or different values of the *z*-component of its angular momentum? Explain.

    b. What angular-momentum quantum number characterizes these atoms? Explain.

41.4 Multielectron Atoms

41.5 The Periodic Table of the Elements

12. Do the following figures represent a possible electron configuration of an element? If so:
    i. Identify the element, and
    ii. Determine if this is the ground state or an excited state.
    If not, why not?

a.

b.

c.

13. Do the following electron configurations represent a possible state of an element? If so:
    i. Identify the element, and
    ii. Determine if this is the ground state or an excited state.
    If not, why not?

    a. $1s^2 2s^2 2p^6 3s^2$

    b. $1s^2 2s^2 2p^7 3s$

    c. $1s^2 2s^2 2p^6 3s^2 3p 3d$

14. Why is the section of the periodic table labeled as "transition elements" exactly 10 elements wide in all rows?

41.6 Excited States and Spectra

15. The figure shows the energy levels of a hypothetical atom.

    a. What is the atom's ionization energy?

    b. In the space below, draw the energy-level diagram as it would appear if the ground state were chosen as the zero of energy. Label each level and the ionization limit with the appropriate energy.

eV
— — — — — — — — — — — — — — — — — — 0
                                      7d   −1
                              7p   −2
                7s   −4
                              6p   −7
        6s   −9
    Ground state

16. The figure shows the energy levels of a hypothetical atom.

    a. What *minimum* kinetic energy (in eV) must an electron have to collisionally excite this atom and cause the emission of a 620 nm photon? Explain.

    b. Can an electron with $K = 6$ eV cause the emission of 620 nm light? If so, what is the final kinetic energy of the electron? If not, why not?

    c. Can a 6 eV photon cause the emission of 620 nm light from this atom? Why or why not?

    d. Can 7 eV photons cause the emission of 620 nm light from these atoms? Why or why not?

eV

7 — — — — — — — — — — — — — — — $(n+1)p$

5 — — — — — $(n+1)s$
                            620 nm
3 — — — — — — — — — — — — — — — np

0 — — — ns
    Ground state

17. Seven possible transitions are identified on the energy-level diagram. For each, is this an allowed transition?

   i. If allowed, is it an emission or an absorption transition?
   ii. Is the photon infrared, visible, or ultraviolet?
   iii. If not allowed, why not?

Transition 1:

Transition 2:

Transition 3:

Transition 4:

Transition 5:

Transition 6:

Transition 7:

41.7 Lifetimes of Excited States

18. A sample of atoms is excited at $t = 0$ ns. The graph shows the number of excited atoms as a function of time after $t = 0$ ns. Each atom decays back to the ground state by emitting a photon.

a. How many atoms were initially excited?

b. At what time have half of the excited atoms decayed?

c. How many photons are emitted in the first 20 ns?

d. At what time will there be 25 atoms remaining in the excited state? Explain.

e. As best as you can determine from this graph, what is the lifetime of this excited state? Explain how you determined it.

19. At $t = 0$ ns, 1000 atoms are excited to a state that has a lifetime of 5 ns.

a. Sketch a graph of the number of excited atoms N_{exc} still present at time t. Include a numerical scale on the vertical axis.

b. Sketch a graph of the total number of photons $N_{photons}$ that have been emitted by time t. Include a numerical scale on the vertical axis.

41.8 Stimulated Emission and Lasers

20. A photon with energy 2.0 eV is incident on an atom in the *p*-state. Does the atom undergo an absorption transition, a stimulated emission transition, or neither? Explain.

21. A glass tube contains 2×10^{11} atoms, some of which are in the ground state and some of which are excited. The populations are shown for the atom's three energy levels. Is it possible for these atoms to be a laser? If so, on which transition would laser action occur? If not, why not?

Level 3
s-state
$N_3 = 8 \times 10^{10}$

Level 2
p-state
$N_2 = 2 \times 10^{10}$

$N_1 = 10 \times 10^{10}$
s-state
Level 1

42 Nuclear Physics

42.1 Nuclear Structure

1. Consider the atoms ^{16}O, ^{18}O, ^{18}F, ^{18}Ne, and ^{20}Ne. Some of the questions about these atoms may have more than one answer. Give all answers that apply.

   a. Which atoms are isotopes? _____

   b. Which atoms are isobars? _____

   c. Which atoms have the same chemical properties? _____

   d. Which atoms have the same number of neutrons? _____

   e. Which atoms have the same number of valence electrons? _____

   f. Rank in order, from largest to smallest, the radii of the nuclei of these atoms.

2. Here are several stable or very-long-lived nuclei:

^{58}Fe		^{58}Ni
^{60}Fe	^{60}Co	^{60}Ni
^{62}Fe		^{62}Ni

   a. Which 2 nuclei are isotopes of ^{60}Fe? _____

   b. Which 2 nuclei are isobars of ^{60}Ni? _____

3. The atomic ion S^- has (circle one):

   i. Lost a proton.   iii. Either a or b.

   ii. Gained an extra electron.   iv. Turned a proton into a neutron.

   Explain your choice.

42.2 Nuclear Stability

4. a. Is there a ^{30}Li ($Z = 3$) nucleus? If so, is it stable or radioactive? If not, why not?

   b. Is there a ^{134}U ($Z = 92$) nucleus? If so, is it stable or radioactive? If not, why not?

5. a. Is the total binding energy of a nucleus with $A = 200$ more than, less than, or equal to the binding energy of a nucleus with $A = 60$? Explain.

   b. Is a nucleus with $A = 200$ more tightly bound, less tightly bound, or bound equally tightly as a nucleus with $A = 60$? Explain.

6. Rounding slightly, the nucleus ^{3}He has a binding energy of 2.5 MeV/nucleon and the nucleus ^{6}Li has a binding energy of 5 MeV/nucleon.

   a. What is the total binding energy of ^{3}He? _____

   b. What is the total binding energy of ^{6}Li? _____

   c. Is it energetically possible for two ^{3}He nuclei to join or fuse together into a ^{6}Li nucleus? Explain.

   d. Is it energetically possible for a ^{6}Li nucleus to split or fission into two ^{3}He nuclei? Explain.

42.3 The Strong Force

42.4 The Shell Model

7. a. How do we know the strong force exists?

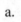

b. How do we know the strong force is short range?

8. For each nuclear energy-level diagram, state whether this represents a nuclear ground state, an excited nuclear state, or an impossible nucleus.

a.

b.

c.

9. Draw energy-level diagrams showing the nucleons in ^{6}Li and ^{7}Li.

a.

^{6}Li

b.

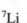

^{7}Li

42.5 Radiation and Radioactivity

10. What is the half-life of this nucleus?

11. A radioactive sample has a half-life of 10 s. 10,000 nuclei are present at $t = 20$ s.

    a. How many nuclei were there at $t = 0$ s? _____

    b. How many nuclei will there be at $t = 40$ s? _____

12. Nucleus A decays into nucleus B with a half-life of 10 s. At $t = 0$ s there are 1000 A nuclei and no B nuclei. At what time will there be 750 B nuclei?

13. A radioactive sample with a half-life of 1.0 min is prepared at $t = 0$ min. One particular nucleus has not decayed at $t = 15$ min. What is the probability that this nucleus will decay sometime between $t = 15$ min and $t = 16$ min?

14. You have four radioactive samples:

Sample	Moles	Activity (Bq)
A	1	100
B	10	100
C	100	100
D	100	1000

    Rank in order, from largest to smallest, the half-lives of these four samples.

    Order:

    Explanation:

42.6 Nuclear Decay Mechanisms

15. Identify the unknown X in the following decays:

    a. $^{222}Rn(Z = 86) \rightarrow {}^{218}Po(Z = 84) + X$ X = _____

    b. $^{228}Ra(Z = 88) \rightarrow {}^{228}Ac(Z = 89) + X$ X = _____

    c. $^{140}Xe(Z = 54) \rightarrow {}^{140}Cs(Z = 55) + X$ X = _____

    d. $^{64}Cu(Z = 29) \rightarrow {}^{64}Ni(Z = 28) + X$ X = _____

    e. $^{18}F(Z = 9) + X \rightarrow {}^{18}O\ (Z = 8)$ X = _____

16. Are the following decays possible? If not, why not?

    a. $^{232}Th(Z = 90) \rightarrow {}^{236}U(Z = 92) + \alpha$

    b. $^{238}Pu(Z = 94) \rightarrow {}^{236}U(Z = 92) + \alpha$

    c. $^{24}Na(Z = 11) \rightarrow {}^{24}Na(Z = 11) + \gamma$

    d. $^{33}P(Z = 15) \rightarrow {}^{32}S(Z = 16) + e^-$

17. An alpha particle is ejected from a nucleus with 6 MeV of kinetic energy. Was the alpha particle's kinetic energy inside the nucleus more than, less than, or equal to 6 MeV? Explain.

18. Three alpha particle energy levels in a nuclear potential energy well are shown below.

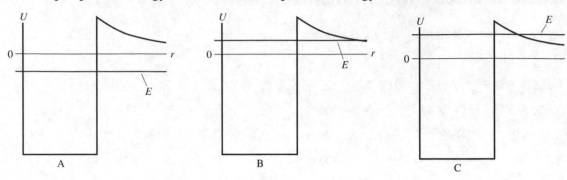

a. Which can undergo alpha decay? _____

b. Of those that decay, which has the longest half-life? _____

c. Of those that decay, which ejects the alpha particle with the largest kinetic energy? _____

19. What kind of decay, if any, can occur for these nuclei?

a.   b.   c.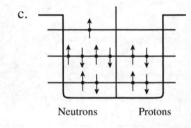

_____   _____   _____

20. Part of the ^{236}U decay series is shown.

a. Complete the labeling of atomic numbers and elements on the bottom edge.

b. Label each arrow to show the type of decay.

42.7 Biological Applications of Nuclear Physics

21. A and B are apples. Apple A is strongly irradiated by nuclear radiation for 1 hour. Apple B is not irradiated. Afterward, in what ways are apples A and B different?

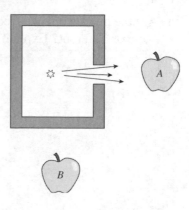

22. Four radiation doses are as follows.

Dose	Gray	RBE
A	10	1
B	20	1
C	10	2
D	20	2

a. Rank in order, from largest to smallest, the amount of energy delivered by these four doses.

b. Rank in order, from largest to smallest, the biological damage caused by these four doses.

23. Consider the following three decays:

$$^{212}\text{Po} \rightarrow {}^{208}\text{Pb} + \alpha$$

$$^{137}\text{Cs} \rightarrow {}^{137}\text{Ba} + e^- + \gamma$$

$$^{90}\text{Sr} \rightarrow {}^{90}\text{Y} + e^-$$

Which of the three isotopes, ^{212}Po, ^{137}Cs, or ^{90}Sr, would be most useful as a tracer? Why?

24. The magnetic field through a patient increases from left to right.

    Increasing field strength

    a. Are the nmr resonance frequencies at points 1, 2, and 3 the same or different? Explain.

    b. If they are different, which is the highest resonance frequency?

DYNAMICS WORKSHEET   Name _____   Problem _____

MODEL Make simplifying assumptions.

VISUALIZE

- Draw a picture. Show important points in the motion.
- Establish a coordinate system. Define symbols.
- List knowns. Identify what you're trying to find.

- Draw a motion diagram.
- Identify forces and interactions.
- Draw a free-body diagram.

Known

Find

SOLVE

Start with Newton's second law in component form, adding other information as needed to solve the problem.

ASSESS

Have you answered the question?
Do you have correct units, signs, and significant figures?
Is your answer reasonable?

DYNAMICS WORKSHEET Name _____ Problem _____

MODEL Make simplifying assumptions.

VISUALIZE
- Draw a picture. Show important points in the motion.
- Establish a coordinate system. Define symbols.
- List knowns. Identify what you're trying to find.
- Draw a motion diagram.
- Identify forces and interactions.
- Draw a free-body diagram.

Known

Find

SOLVE
Start with Newton's second law in component form, adding other information as needed to solve the problem.

ASSESS
Have you answered the question?
Do you have correct units, signs, and significant figures?
Is your answer reasonable?

DYNAMICS WORKSHEET   Name _____   Problem _____

MODEL Make simplifying assumptions.

VISUALIZE

- Draw a picture. Show important points in the motion.
- Establish a coordinate system. Define symbols.
- List knowns. Identify what you're trying to find.

- Draw a motion diagram.
- Identify forces and interactions.
- Draw a free-body diagram.

Known

Find

SOLVE
Start with Newton's second law in component form, adding other information as needed to solve the problem.

ASSESS
Have you answered the question?
Do you have correct units, signs, and significant figures?
Is your answer reasonable?

DYNAMICS WORKSHEET Name _____ Problem _____

MODEL Make simplifying assumptions.

VISUALIZE
- Draw a picture. Show important points in the motion.
- Establish a coordinate system. Define symbols.
- List knowns. Identify what you're trying to find.
- Draw a motion diagram.
- Identify forces and interactions.
- Draw a free-body diagram.

Known

Find

SOLVE
Start with Newton's second law in component form, adding other information as needed to solve the problem.

ASSESS
Have you answered the question?
Do you have correct units, signs, and significant figures?
Is your answer reasonable?

ENERGY WORKSHEET Name _____ Problem _____

MODEL Make simplifying assumptions.

VISUALIZE

- Draw a before-and-after picture.
- Establish a coordinate system. Define symbols.

- Draw an energy bar chart.
- List knowns. Identify what you're trying to find.

Known

Find

What is the system? _____

Potential energies? _____

Nonconservative forces? _____

External forces? _____

Is mechanical energy conserved? _____

$$K_i \quad + \quad U_i \quad + \quad W_{ext} \quad = \quad K_f \quad + \quad U_f \quad + \quad \Delta E_{th}$$

SOLVE

Start with conservation of energy or the energy principle, adding other information as needed to solve the problem.

ASSESS

Have you answered the question?
Do you have correct units, signs, and significant figures?
Is your answer reasonable?

ENERGY WORKSHEET Name _____ Problem _____

MODEL Make simplifying assumptions.

VISUALIZE

- Draw a before-and-after picture.
- Establish a coordinate system. Define symbols.

- Draw an energy bar chart.
- List knowns. Identify what you're trying to find.

Known

Find

What is the system? _____

Potential energies? _____

Nonconservative forces? _____

External forces? _____

Is mechanical energy conserved? _____

$$K_i + U_i + W_{ext} = K_f + U_f + \Delta E_{th}$$

SOLVE

Start with conservation of energy or the energy principle, adding other information as needed to solve the problem.

ASSESS

Have you answered the question?
Do you have correct units, signs, and significant figures?
Is your answer reasonable?

ENERGY WORKSHEET   Name _____   Problem _____

MODEL Make simplifying assumptions.

VISUALIZE

- Draw a before-and-after picture.
- Establish a coordinate system. Define symbols.

- Draw an energy bar chart.
- List knowns. Identify what you're trying to find.

Known

Find

What is the system? _____

Potential energies? _____

Nonconservative forces? _____

External forces? _____

Is mechanical energy conserved? _____

$$K_i \quad + \quad U_i \quad + \quad W_{ext} \quad = \quad K_f \quad + \quad U_f \quad + \quad \Delta E_{th}$$

SOLVE

Start with conservation of energy or the energy principle, adding other information as needed to solve the problem.

ASSESS

Have you answered the question?
Do you have correct units, signs, and significant figures?
Is your answer reasonable?

ENERGY WORKSHEET   Name _____   Problem _____

MODEL Make simplifying assumptions.

VISUALIZE
- Draw a before-and-after picture.
- Establish a coordinate system. Define symbols.

- Draw an energy bar chart.
- List knowns. Identify what you're trying to find.

Known

Find

What is the system? _____

Potential energies? _____

Nonconservative forces? _____

External forces? _____

Is mechanical energy conserved? _____

$$K_i \quad + \quad U_i \quad + \quad W_{ext} \quad = \quad K_f \quad + \quad U_f \quad + \quad \Delta E_{th}$$

SOLVE
Start with conservation of energy or the energy principle, adding other information as needed to solve the problem.

ASSESS
Have you answered the question?
Do you have correct units, signs, and significant figures?
Is your answer reasonable?

MOMENTUM WORKSHEET   Name _____   Problem _____

MODEL   Make simplifying assumptions.

VISUALIZE
- Draw a before-and-after picture.
- Establish a coordinate system. Define symbols.
- Draw a momentum bar chart.
- List knowns. Identify what you're trying to find.

Known

Find

- What is the system? _____
- What forces exert impulses on the system? _____
- Is the system's momentum conserved during part or all of the problem?

  If so, during which part? _____

$$P_{ix} \quad + \quad J_x \quad = \quad P_{fx}$$

SOLVE
Start with conservation of momentum or the momentum principle, using Newton's laws or kinematics as needed.

ASSESS
Have you answered the question?
Do you have correct units, signs, and significant figures?
Is your answer reasonable?

MOMENTUM WORKSHEET Name _____ Problem _____

MODEL Make simplifying assumptions.

VISUALIZE

- Draw a before-and-after picture.
- Establish a coordinate system. Define symbols.

- Draw a momentum bar chart.
- List knowns. Identify what you're trying to find.

Known

Find

- What is the system? _____
- What forces exert impulses on the system? _____
- Is the system's momentum conserved during part or all of the problem?

  If so, during which part? _____

  _____

$$P_{ix} \quad + \quad J_x \quad = \quad P_{fx}$$

SOLVE

Start with conservation of momentum or the momentum principle, using Newton's laws or kinematics as needed.

ASSESS

Have you answered the question?
Do you have correct units, signs, and significant figures?
Is your answer reasonable?

MOMENTUM WORKSHEET Name _____ Problem _____

MODEL Make simplifying assumptions.

VISUALIZE

- Draw a before-and-after picture.
- Establish a coordinate system. Define symbols.
- Draw a momentum bar chart.
- List knowns. Identify what you're trying to find.

Known

Find

- What is the system? _____
- What forces exert impulses on the system? _____
- Is the system's momentum conserved during part or all of the problem?

  If so, during which part? _____

  _____

$$P_{ix} + J_x = P_{fx}$$

SOLVE

Start with conservation of momentum or the momentum principle, using Newton's laws or kinematics as needed.

ASSESS

Have you answered the question?
Do you have correct units, signs, and significant figures?
Is your answer reasonable?

MOMENTUM WORKSHEET Name _____ Problem _____

MODEL Make simplifying assumptions.

VISUALIZE

- Draw a before-and-after picture.
- Establish a coordinate system. Define symbols.

- Draw a momentum bar chart.
- List knowns. Identify what you're trying to find.

Known

Find

- What is the system? _____
- What forces exert impulses on the system? _____
- Is the system's momentum conserved during part or all of the problem?

  If so, during which part? _____

$$P_{ix} \quad + \quad J_x \quad = \quad P_{fx}$$

SOLVE

Start with conservation of momentum or the momentum principle, using Newton's laws or kinematics as needed.

ASSESS

Have you answered the question?
Do you have correct units, signs, and significant figures?
Is your answer reasonable?